Frontiers in Reproductive Science

(Volume 1)

(Reproductive Biology, Physiology and Biochemistry of Male Bats)

Edited by

Edith Arenas-Ríos

Department of Biology of Reproduction, Universidad Autónoma Metropolitana, Iztapalapa, 186, St. Rafael Atlixco, Vicentina, 09340 México City, México

General:

1. Any dispute or claim arising out of or in connection with this License Agreement or the Work (including non-contractual disputes or claims) will be governed by and construed in accordance with the laws of the U.A.E. as applied in the Emirate of Dubai. Each party agrees that the courts of the Emirate of Dubai shall have exclusive jurisdiction to settle any dispute or claim arising out of or in connection with this License Agreement or the Work (including non-contractual disputes or claims).
2. Your rights under this License Agreement will automatically terminate without notice and without the need for a court order if at any point you breach any terms of this License Agreement. In no event will any delay or failure by Bentham Science Publishers in enforcing your compliance with this License Agreement constitute a waiver of any of its rights.
3. You acknowledge that you have read this License Agreement, and agree to be bound by its terms and conditions. To the extent that any other terms and conditions presented on any website of Bentham Science Publishers conflict with, or are inconsistent with, the terms and conditions set out in this License Agreement, you acknowledge that the terms and conditions set out in this License Agreement shall prevail.

Bentham Science Publishers Ltd.
Executive Suite Y - 2
PO Box 7917, Saif Zone
Sharjah, U.A.E.
Email: subscriptions@benthamscience.org

BENTHAM SCIENCE

CONTENTS

PREACE ... i

LIST OF CONTRIBUTORS ... ii

DEDICATION ... iii

CHAPTER 1 ULTRASTRUCTURE OF SPERMATOGENESIS AND SPERM OF BATS ... 1

Mateus Rodrigues Beguelini, Sebastião Roberto Taboga and *Eliana Morielle-Versute*

SPERMATOGENESIS ... 2

ULTRASTRUCTURE OF SPERMATOGENESIS IN BATS ... 2

Proliferation Phase ... 2

Meiotic Phase ... 4

Spermiogenesis ... 5

Golgi Phase ... 7

Cap Phase ... 10

Acrosome Phase ... 10

Maturation Phase ... 11

ULTRASTRUCTURE OF BAT SPERMATOZOA ... 12

ULTRASTRUCTURE OF THE SUPPORTING CELLS OF BATS ... 14

Sertoli Cells ... 14

Leydig Cells ... 15

PERSPECTIVES ... 16

CONFLICT OF INTEREST ... 16

ACKNOWLEDGEMENTS ... 16

REFERENCES ... 16

CHAPTER 2 MALE REPRODUCTIVE HORMONAL REGULATION IN BATS AND THE VARIATIONS ASSOCIATED WITH TESTICULAR REGRESSION ... 20

Mateus Rodrigues Beguelini, Sebastião Roberto Taboga and *Eliana Morielle-Versute*

MAMMALIAN HORMONAL REGULATION ... 21

The Hypothalamic-Pituitary-Testicular Axis (HPT) ... 21

The Paracrine Regulation of the Testes ... 22

Regulation of Secondary Reproductive Organs ... 24

Epididymal Regulation ... 24

Prostatic Regulation ... 25

HORMONAL REGULATION IN BATS ... 26

Hormonal Regulation of Testicular Regression in Hibernating Bats ... 28

Hormonal Regulation of Testicular Regression in Nonhibernating Bats ... 29

PERSPECTIVES ... 32

CONFLICT OF INTEREST ... 32

ACKNOWLEDGEMENTS ... 32

REFERENCES ... 32

CHAPTER 3 RELATIONSHIP BETWEEN SUBCUTANEOUS FAT AND REPRODUCTIVE ACTIVITY IN MALES OF LEPTONYCTERIS yerbabuenae IN LOS LAGUITOS CAVE, CHIAPAS, MEXICO ... 36

Matías Martínez-Coronel, Yolanda Hortelano-Moncada, Violeta Corral and *Laura Cuevas*

STUDY AREA ... 38

METHODOLOGY ... 38

DATA ANALYSIS ... 39

RESULTS ... 39

Reproductive Phenology ... 39

Relationships between Body Mass, Testicular Mass and Fat Index (FI) ... 40
Dorsal Patch Development ... 40
Breeding Colony ... 43
DISCUSSION ... 44
CONCLUSION ... 46
CONFLICT OF INTEREST ... 46
ACKNOWLEDGEMENTS ... 46
REFERENCES ... 46

CHAPTER 4 EVIDENCE OF APOPTOSIS IN THE TESTICULAR BLOOD VESSELS OF *CORYNORHINUS mexicanus* BATS CHARACTERIZED BY SEASONAL REPRODUCTION ... 49
Miguel Angel León-Galván, Ahiezer Rodríguez-Tobón and *Rosa María Vigueras-Villaseñor*
INTRODUCTION ... 50
REPRODUCTIVE PATTERNS IN BATS ... 50
EFFECT OF PHOTOPERIOD AND TEMPERATURE ON TESTICULAR FUNCTION ... 52
SPERMATOGENESIS IN BATS ... 53
MORPHOPHYSIOLOGICAL CHANGES IN MAMMAL TESTICLES ACCORDING TO SEASONAL REPRODUCTION ... 55
PARTICIPATION OF APOPTOSIS IN REGULATING MALE GERMINAL CELLS ... 56
APOPTOSIS IN THE TESTICLES OF MAMMALS WITH SEASONAL REPRODUCTION ... 58
TESTICULAR CYCLE OF *CORYNORHINUS MEXICANUS* BAT ... 59
APOPTOSIS IN THE TESTICULAR BLOOD VESSELS OF *CORYNORHINUS MEXICANUS* ... 62
CONFLICT OF INTEREST ... 66
ACKNOWLEDGEMENTS ... 67
REFERENCES ... 67

CHAPTER 5 EPIDIDYMAL SPERM MATURATION IN BATS ... 74
Edith Arenas-Ríos, Ahiezer Rodríguez-Tobón and *Miguel Ángel León-Galván*
INTRODUCTION ... 74
REPRODUCTION IN TEMPERATE ZONE BATS ... 75
Male Reproductive Patterns ... 76
Temporal Asynchrony between Male Reproductive Functions and Prolonged Sperm Storage 77
EPIDIDYMAL SPERM MATURATION IN BATS ... 78
Morphology and Histology of the Epididymis ... 78
Incorporation of Proteins into Sperm and the Participation of Epididymisomes in the Maturation Process ... 81
Modification of Sperm Proteins During Maturation ... 84
Protein Glycosylation ... 84
Protein Phosphorylation ... 86
REGULATION OF THE MATURATION PROCESS ... 87
Regulation by Androgens ... 87
Regulation by Reactive Oxygen Species ... 89
THE *CORYNORHINUS MEXICANUS* BAT AS A MODEL FOR STUDYING THE ESM PROCESS ... 91
CONCLUSION AND PERSPECTIVES ... 92
CONFLICT OF INTEREST ... 93
ACKNOWLEDGEMENTS ... 93
REFERENCES ... 93

CHAPTER 6 **MECHANISMS OF SPERM-STORAGE IN FEMALE REPRODUCTIVE TRACT IN BATS** 103
Amitabh Krishna and *Vikas Kumar Roy*
INTRODUCTION 103
SPERM-STORAGE SITE 104
MECHANISM OF SPERM-STORAGE 105
Low Body Temperature Facilitate Sperm-storage 106
Unique Relationship between Spermatozoa and Storage Organ 106
Hormonal Regulation of Sperm-Storage 107
Nutritional Support and Metabolic Changes 108
Capacitation and Sperm-Storage 109
Studies on Bats, Scotophilus heathi: 109
SIGNIFICANCE OF SPERM STORAGE 112
CONCLUSION 112
ABBREVIATIONS 112
CONFLICT OF INTEREST 113
ACKNOWLEDGEMENTS 113
REFERENCES 113

CHAPTER 7 **SPERM COMPETITION AND CRYPTIC FEMALE CHOICE IN BATS** 119
C. Ruth Archer, Teri J. Orr and *David J. Hosken*
SEXUAL SELECTION 119
POST-COPULATORY SEXUAL SELECTION: COMPETING SPERM AND CRYPTIC CHOICE 120
BAT REPRODUCTION 122
THE EVOLUTIONARY CONSEQUENCES OF SPERM COMPETITION IN BATS 125
Sperm Competition and Sperm Numbers 125
Sperm Size, Structure and Swimming Speed 126
Seminal Fluid and Copulatory Plugs 130
Genital Morphology 134
Behavioural Adaptations 136
Cryptic Female Choice 136
FUTURE DIRECTIONS AND CHALLENGES 137
CONFLICT OF INTEREST 138
ACKNOWLEDGEMENTS 138
REFERENCES 138

GLOSSARY 148

SUBJECT INDEX 151

PREFACE

The knowledge we have today about reproduction in mammals is well advanced, the morphological aspects, endocrine, biochemical, molecular as well as regulatory mechanisms and signal transduction related to reproduction, can be manipulated to support assisted animal reproduction.

Thus, the scientists focusing on the biology of reproduction have been able to clone and freeze gametes in germ cell transplantation, with worth-mentioning advances in basic research.

A review of books on reproduction in mammals, do not provide extensive information; however, books published on reproduction in bats have opened to us the fascinating world of reproduction, which delineates some of the paradigms established for the generality of mammals. Books on reproduction in bats, contain information covering aspects of reproductive ecophysiology, morphology and little about reproductive physiology, but none is focused on the physiological and biochemical aspects of reproduction in bats, as they were published at a time when information about these species was not enough to explain these processes. However, recently, many articles have been published that intend to explain the extraordinary physiology and biochemistry of reproductive processes in bats.

Although the literature on reproduction in bats is limited as compared to reproduction in mammals, this book only contains seven chapters on advances in biology, physiology and biochemistry of reproduction in male bats and its sperm cells.

Edith Arenas Ríos
Department of Biology of Reproduction
Universidad Autónoma Metropolitana, Iztapalapa
186, St. Rafael Atlixco
Vicentina, 09340 México City
México

List of Contributors

Ahiezer Rodríguez-Tobón	Departamento de Biología de la Reproducción, Universidad Autónoma Metropolitana - Iztapalapa, Ciudad de México, México
Amitabh Krishna	Department of Zoology, Institute of Sciences, Banaras Hindu University, Varanasi, India
C. Ruth Archer	Centre for Ecology and Conservation, College of Life and Environmental Sciences, University of Exeter, Penryn, UK
David J. Hosken	Centre for Ecology and Conservation, College of Life and Environmental Sciences, University of Exeter, Tremough Campus, TR10 9EZ, Penryn, UK
Edith Arenas-Ríos	Departamento de Biología de la Reproducción, Universidad Autónoma Metropolitana - Iztapalapa, Ciudad de México, México
Eliana Morielle-Versute	Department of Zoology and Botany, UNESP – Univ. Estadual Paulista, São Paulo, Brazil
Laura Cuevas	Departamento de Biología, UAM-Iztapalapa, Ciudad de México, México
Mateus Rodrigues Beguelini	Center of Biological and Health Sciences, UFOB – Universidade Federal do Oeste da Bahia, Bahia, Brazil
Matías Martínez Coronel	Departamento de Biología, UAM-Iztapalapa, Ciudad de México, México
Miguel Ángel León-Galván	Departamento de Biología de la Reproducción, Universidad Autónoma Metropolitana - Iztapalapa, Ciudad de México, México
Rosa María Vigueras-Villaseñor	Laboratorio de Biología de la Reproducción, Instituto Nacional de Pediatría, Ciudad de México, México
Sebastião Roberto Taboga	Department of Biology, UNESP – Univ. Estadual Paulista, São Paulo, Brazil
Teri J. Orr	Department of Biology, University of Utah, Salt Lake City, UT, USA
Vikas Kumar Roy	Department of Zoology, School of Life Sciences, Mizoram University, Aizawl, India
Violeta Corral	Departamento de Biología, UAM-Iztapalapa, Ciudad de México, México
Yolanda Hortelano-Moncada	Departamento de Zoología, Instituto de Biología, Universidad Nacional Autónoma de México, Ciudad de México, México

DEDICATION

This book is dedicated to the two people who showed me the wonderful world of bats: Drs. Miguel Angel León Galvan and Ricardo López Wilchis; the person who showed me the wonderful world of Science: Dr. Adolfo Rosado García and the person who showed me that besides bats and science, there is another wonderful world: My husband Mario Gabriel Hidalgo.

CHAPTER 1

Ultrastructure of Spermatogenesis and Sperm of Bats

Mateus Rodrigues Beguelini[1,*], **Sebastião Roberto Taboga**[2] and **Eliana Morielle-Versute**[3]

[1] *Center of Biological and Health Sciences, UFOB – Universidade Federal do Oeste da Bahia, Barreiras, Bahia, 47808-021, Brazil*

[2] *Department of Biology, UNESP – Univ. Estadual Paulista, São José do Rio Preto, São Paulo,15054-000, Brazil*

[3] *Department of Zoology and Botany, UNESP – Univ. Estadual Paulista, São José do Rio Preto, São Paulo, 15054-000, Brazil*

Abstract: Spermatogenesis is a highly regulated and synchronized process of cellular division and differentiation whereby spermatogonia proliferate, meiotically divide, and gradually differentiate into highly specialized haploid cells, known as spermatozoa. Its main events are generally similar in most mammals, and it may be divided into three phases: the proliferation phase, the meiotic phase, and spermiogenesis. The proliferation phase corresponds to the process of stem cell renewal, where spermatogonia mitotically divide to both replenish the stem cell pool and to originate cells that are subject to further differentiation. The meiotic phase is the process in which spermatocytes undergo meiotic divisions that give rise to haploid spermatids. Moreover, spermiogenesis is the process in which a round spermatid differentiates into a mature spermatozoon that is capable of motility and fertilization. The process of spermatogonial differentiation in bats is relatively similar to that found in primates, with three main types of spermatogonia: A_d, A_p, and B. Meiotic divisions proceed similarly to those of most mammals, and spermiogenesis can be divided into 9 to 16 steps. Despite these similarities, some species-specific variations are observed. Bats present three different processes for the formation of the acrosome, and the ultrastructure of spermatozoa has been found to have unique characteristics, including many wavy acrosomal projections on the acrosome; surface of the family Molossidae, an extraordinarily large head with accordion-like folds of the Noctilionidae, differences in the degree of development of the outer dense fibers, and the presence/absence of a perforatorium.

Keywords: Acrosome, Bat, Chiroptera, Golgi complex, Molossidae, Perforatorium, Phyllostomidae, Sperm, Spermatogenesis, Spermiogenesis, Testis, Ultrastructure, Vespertilionidae.

* **Corresponding author Mateus Rodrigues Beguelini:** Rua Prof. José Seabra de Lemos nº 316, Recanto dos Pássaros, 47808-021, Barreiras, Bahia, Brazil; Tel: +55 77 36143215; E-mail: mateus.beguelini@ufob.edu.br

Edith Arenas-Ríos (Ed.)

SPERMATOGENESIS

Spermatogenesis is a highly regulated and synchronized process of cellular division and differentiation whereby stem cells (spermatogonia) proliferate, meiotically divide, and gradually differentiate into highly specialized haploid cells, spermatozoa [1-2]. This differentiation process has attracted the attention of several researchers, not only because it is one of the most finely regulated cellular process, which involves numerous radical changes in the cell shape, biochemistry, and gene regulation, but also because deviations in the process give rise to variations in the final structure of the sperm produced.

It is well known that morphological variations in the same structure may indicate physiological or operational modifications, which, in the case of sperm, seem to affect the capacities of survival, movement, and fertilization. This has led some researchers to propose that the structure (mainly the ultrastructure) of sperm is species-specific or, at least, similar only between closely related species. Based on this, several studies have used the ultrastructure of sperm as an important tool in the identification and classification of taxa of various groups of organisms as well as in phylogenetic analysis [3 - 6].

The main events of the spermatogenesis process are generally similar in most mammals, and spermatogenesis may be divided into three functionally distinct phases: the proliferation phase, the meiotic phase, and spermiogenesis [1, 7].

The proliferation phase corresponds to the process of stem cell renewal, where spermatogonia mitotically divide to both replenish the stem cell pool and to originate cells that are subject to further differentiation, producing spermatocytes. The meiotic phase is the process in which spermatocytes undergo meiotic divisions that give rise to haploid spermatids. Moreover, the last and longest phase of spermatogenesis corresponds to spermiogenesis, the process in which a round spermatid differentiates into a mature spermatozoon that is capable of motility and fertilization [7, 8].

Despite the similarities in the process, variations among and within orders, families, and taxa are frequently observed [9 - 11], mainly determined by variations in the final structure of the spermatozoon [12, 13].

ULTRASTRUCTURE OF SPERMATOGENESIS IN BATS

Proliferation Phase

Spermatogenesis begins with a clonal proliferation process of spermatogonia, which mitotically divide to both replenish the stem cell pool and to be subjected to further differentiation (Fig. **1**). Spermatogonia are enclosed in the basal compartment of the seminiferous tubule, attached to the basal lamina and in close relation with Sertoli cells [12].

Several studies have proposed different recognition methods to classify spermatogonia (based on topographic arrangement, the number of connected cells, the cell morphology, *etc.*); however, one of the methods seems to be more frequently applied to bats. This method is mainly based on the nuclear morphology, degree of chromatin condensation, and cellular position within the seminiferous epithelium.

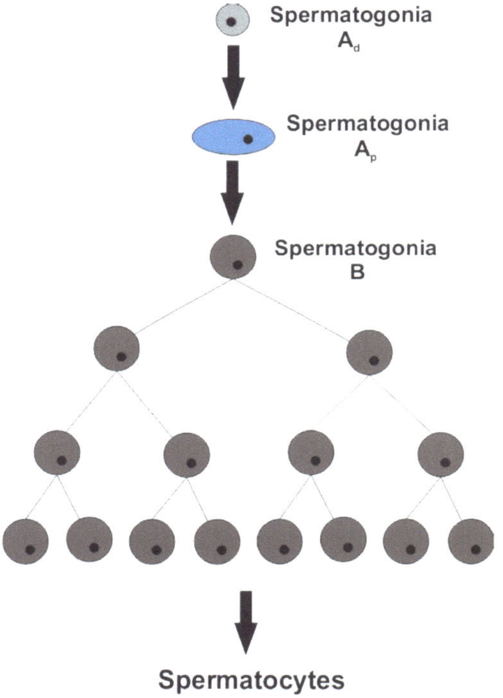

Fig. (1). Schematic representation showing the differentiation process of spermatogonia in bats. Note that the spermatogonial differentiation process begins with the true spermatogonial stem cell, denominated A_d. They can be stimulated to divide to both; to replenish the stem cell pool, forming new A_d spermatogonia; and/or to form cells that will be committed to further differentiation, thus forming the type A_p spermatogonia. The type A_p spermatogonia differentiates in a different cell type, the type B spermatogonia, which divide mitotically, clonally increasing the number of spermatogonia in the seminiferous tubules. At the end of mitotic divisions, these cells release the basal lamina, enter into the adluminal compartment, and differentiate into spermatocytes, entering into meiotic division.

Following this method, Singwi and Lall [14] described three main types of spermatogonia (A, In, and B) in the bat *Rhinopoma kinneari*; McGuckin and Blackshaw [15] described three types (A, In, and B) in *Pteropus poliocephalus*; Saidapur and Patil [16] described five types (A1, A2, A3, In, and B) in *Rousettus leschenaulti*; Moraes *et al.* [17] described three types (A, In, and B) in *Sturnira lilium*; and Beguelini *et al.* [2, 12, 18 - 20] and Bueno *et al.* [21] described three types (A_d, A_p, and B) in seven species of Neotropical bats.

Despite the variation in the number of spermatogonia, three main types are observed in all studies: A_d, A_p and B. Based on this consideration, we propose that the spermatogonial differentiation process in bats begins with the true spermatogonial stem cell, denominated A_d (Fig. **1**). These cells are the rarest type of spermatogonia; they present a compact shape, with little cytoplasm and a condensed nucleus, which indicates a quiescent pattern (Fig. **2A**). They can be stimulated by paracrine factors of Sertoli and/or Leydig cells [22] to divide to both; to replenish the stem cell pool, forming new A_d spermatogonia; and/or to form cells that will be committed to further differentiation, thus forming the type A_p spermatogonia (Fig. **1**).

The type A_p spermatogonia present a more elongated shape, with large cytoplasm full of organelles, and an elongated nucleus with decondensed chromatin, which indicates a more synthetically active pattern (Fig. **2B**). This cell type differentiates in a different cell type, the type B spermatogonia (Fig. **1**).

The type B spermatogonia present a round nucleus with decondensed chromatin and a large cytoplasm (Figs. **2C-D**). These spermatogonia divide mitotically, clonally increasing the number of spermatogonia in the seminiferous tubules. At the end of mitotic divisions, these cells release the basal lamina, enter into the adluminal compartment (Fig. **2D**), and differentiate into spermatocytes, entering into meiotic division.

Meiotic Phase

The meiotic phase corresponds to the process in which spermatocytes undergo meiotic divisions that give rise to haploid spermatids [8]. In bats, they proceed in a manner similar to those of most mammals, with a single round of DNA replication [23], posterior formation of the synaptonemal complexes, gradual chromatin condensation around them, consequent nucleolar disruption (Fig. **3**), and the cytokinesis and formation of haploid spermatids.

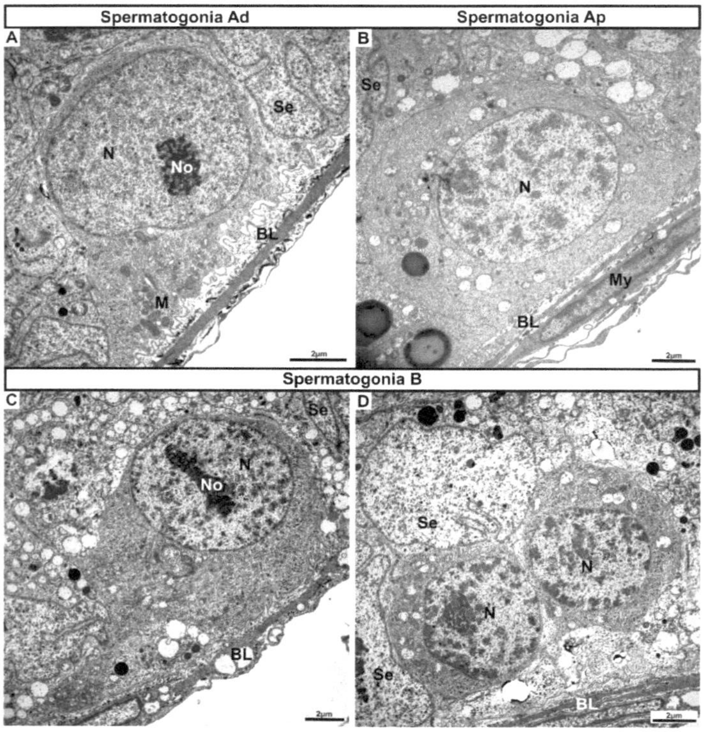

Fig. (2). Transmission electron micrographs of the three types of spermatogonia of bats. **A.** Type A_d spermatogonia of *Platyrrhinus lineatus*. Note the small size of this cell type, its firm adherence to the basal lamina, and the nucleus that occupies most of its cytoplasm. **B.** Type A_p spermatogonia of *Molossus molossus*. Note the more elongated shape of this cell type, with a relatively greater proportion of cytoplasm, the less accentuated adherence to the basal lamina, and the more euchromatic nucleus. **C-D.** Type B spermatogonia of *Platyrrhinus lineatus*. Note that this cell type starts the process of ascension into the adluminal compartment, migrating the nucleus and the cytoplasm to the upstream regions of the basal compartment (**C**), losing contact with the basal lamina (**D**). (BL, basal lamina; M, mitochondria; My, myoid cell; N, nucleus; No, nucleolus; Se, Sertoli cell). Scale bars = 2μm.

Spermiogenesis

The last and longest phase of spermatogenesis corresponds to spermiogenesis, the process in which a round spermatid differentiates into a mature spermatozoon that is capable of motility and fertilization [7]. The spermiogenesis process tends to be species-specific, presenting different characteristics in each species or being similar only between closely related species. This occurs because there are variations in the final ultrastructure of the sperm head of some species, and we know that the more elaborate the structure is, the more complex the process of its formation. Thus, the number of steps in spermatid differentiation in bats varies widely, from 9 to 16 steps [2, 12, 14, 16, 18 - 21, 24 - 26].

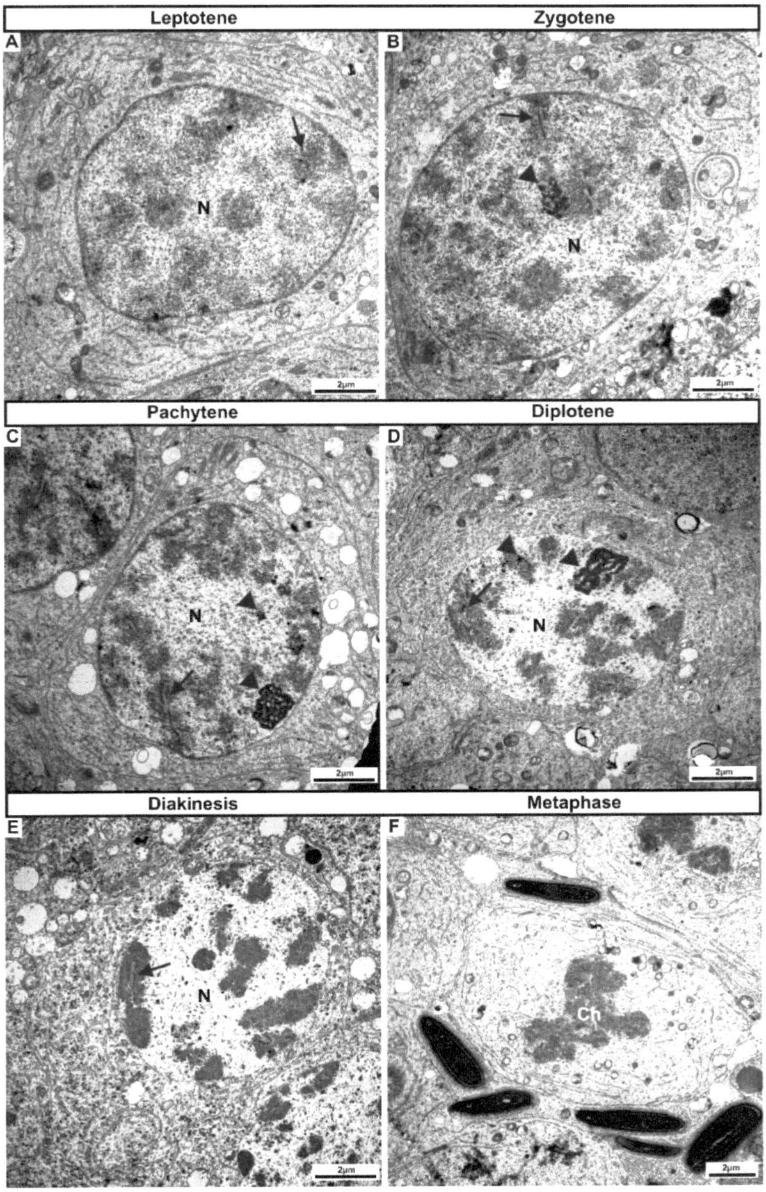

Fig. (3). Electron micrographs of the meiotic cells (meiosis I) of *Molossus molossus* (**A-C**) and *Platyrrhinus linectus* (**D-E**). **A.** Leptotene. **B.** Zygotene. **C.** Pachytene. **D.** Diplotene. **E.** Diakinesis. **F.** Metaphase. Note the increase in the degree of chromatin condensation around the synaptonemal complexes (**A–E**, arrows); the disorganization of the nucleolus from leptotene (**A**) to diplotene (**D**), with a great nucleolar fragmentation in pachytene (arrow-heads); and the structural framework in the middle of the chromosomes in diakinesis (**E**, arrow). (Ch, chromosomes; N, nucleus). Scale bars = 2μm.

Despite the variation in the number of steps, based on morphological changes that occurred during differentiation of spermatids into spermatozoa, the spermiogenesis process can be clearly divided in four phases: the Golgi, the cap, the acrosome, and the maturation phases. In some Neotropical species of bats, these phases have been clearly divided into 12 steps, with the Golgi phase corresponding to steps 1 to 4, the cap phase to steps 5 to 8, the acrosomal phase to steps 9 to 10, and the maturation phase to steps 11 to 12 [2, 12, 19 - 21].

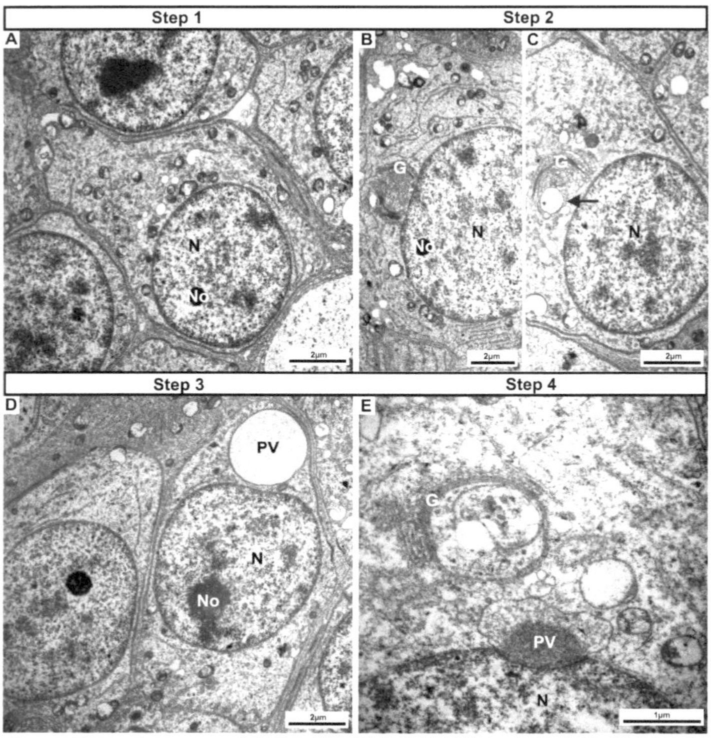

Fig. (4). Electron micrographs of the spermatids of *Molossus molossus* showing the spermiogenesis process (Golgi phase). **A.** Step 1 spermatids. Note that there is little differentiation in this type of spermatid. **B-C.** Step 2 spermatids. Note the considerable development of the Golgi complex and the formation of large proacrosomal vesicles (**C**, arrow). **D.** Step 3 spermatid. Note the large proacrosomal vacuole adjacent to the nucleus. **E.** Step 4 spermatid. Note the attachment of the proacrosomal vacuole to the nuclear envelope, forming a concavity in the nucleus. (G, Golgi complex; N, nucleus; No, nucleolus; PV, proacrosomal vacuole).

Golgi Phase

The Golgi phase is characterized by initiation of the differentiation process, in which round spermatids, newly formed from the second meiotic division, enter into the spermiogenesis process (Fig. **4**). The main processes that occur during this phase are linked to the development of the Golgi complex (Figs. **4B-C**), formation of several proacrosomal vesicles (Fig. **4C**), fusion of the proacrosomal vesicles to form the proacrosomal vacuole (Fig. **4D**), and the attachment of the proacrosomal vacuole to the nucleus to begin the formation of the acrosome (Fig. **4E**). Initiation of the formation of the flagellum also occurs in this phase.

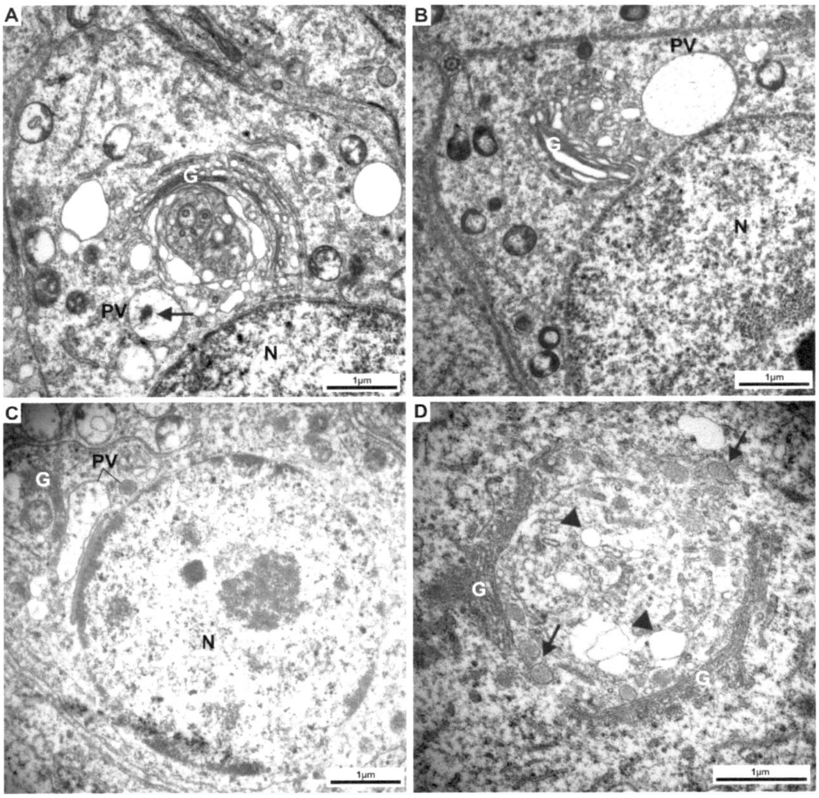

Fig. (5). Electron micrographs showing differences in the content of the proacrosomal vesicles and vacuole in the spermatids of bats. **A.** *Molossus molossus*. Note the presence of an electron-dense granule inside the proacrosomal vacuole (arrow). **B.** *Myotis nigricans*. Note the absence of electron-dense granules, with their proacrosomal vesicles and vacuole being composed only of an electron-lucid matrix. **C-D.** *Platyrrhinus lineatus*. Note that their Golgi complexes produce two different types of proacrosomal vesicles and vacuole – one electron-dense (arrow), which attaches first to the nucleus, and the other electron-lucid, which attaches later (arrowheads) – but with neither of them showing an electron-dense granule inside. (G, Golgi complex; N, nucleus; PV, proacrosomal vacuole).

Although all bats present similar processes during the Golgi phase, some species-specific variations have been observed. Like in most mammals [27 - 29], some species of bats present an electron-dense granule inside the proacrosomal vesicles and vacuole (Fig. **5A**) during the development of the acrosome [2]. However, other species do not have electron-dense granules [19], with their proacrosomal vesicles and vacuole being composed only of an electron-lucid matrix (Fig. **5B**).

In other species [12, 20], we have observed that their Golgi complexes produce two different types of proacrosomal vesicles and vacuoles – one electron-dense, which attaches first to the nucleus, and the other electron-lucid, which attaches later (Figs. **5C-D**) – but with neither of them showing an electron-dense granule inside. This difference can be related to the formation of the perforatorium, where these electron-dense materials may be the basic substance from their mounting. Thus, electron-dense substance are present only in species that have a perforatorium.

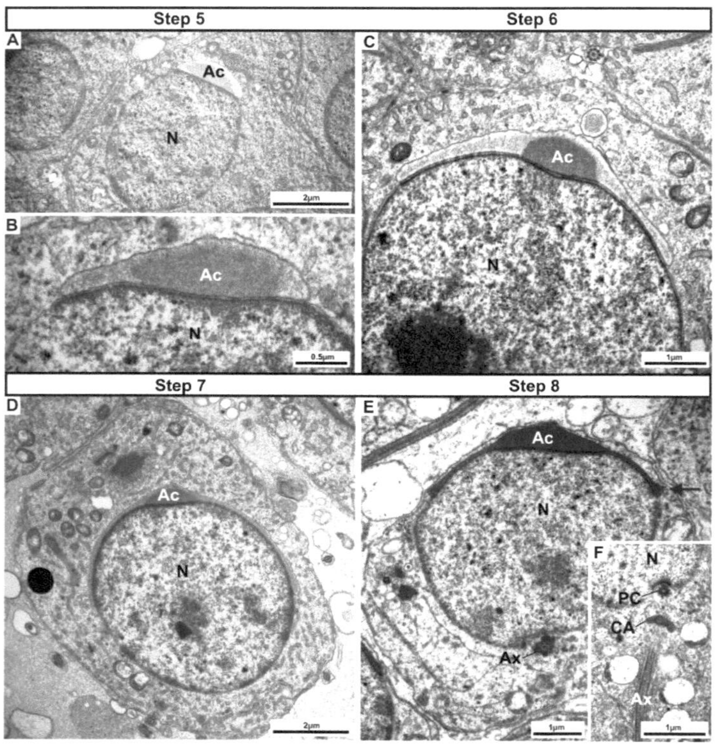

Fig. (6). Electron micrographs of the spermatids of *Platyrrhinus lineatus* (**A-B** and **E**) and *Molossus molossus* (**C-D** and **F**) showing the spermiogenesis process (Cap phase). **A–B.** Step 5 spermatids. Note that the proacrosomal vacuole is firmly attached to the nucleus and beginning to extend, covering the nucleus and forming the acrosomal cap. **C.** Step 6 spermatid. Note the intimate association of the acrosomal cap with the nuclear envelope and two types of components inside, a more electron-dense type and a translucent type. **D.** Step 7 spermatid. Note the expansion of the acrosomal cap, which covers most of the nuclear surface and contains a strong electron- dense matrix. **E-F.** Step 8 spermatids. Note the development of an elevation in the distal region of the acrosome (arrow), the approach of the nucleus to the plasma membrane, and the development of the flagellum. (Ac, acrosomal cap; Ax, axoneme; CA, centriolar adjunct; N, nucleus; PC, proximal centriole).

Cap Phase

In the cap phase (Fig. **6**), the proacrosomal vacuole is flattened against the nucleus (Fig. **6A**) and begins to extend laterally (Fig. **6B**), forming the acrosomal cap and defining the cranial pole of the spermatid (Figs. **6C**-**E**). Similarly, the centrioles migrate to the opposite pole, forming the flagellum and defining the caudal pole of the spermatid (Figs. **6E**-**F**).

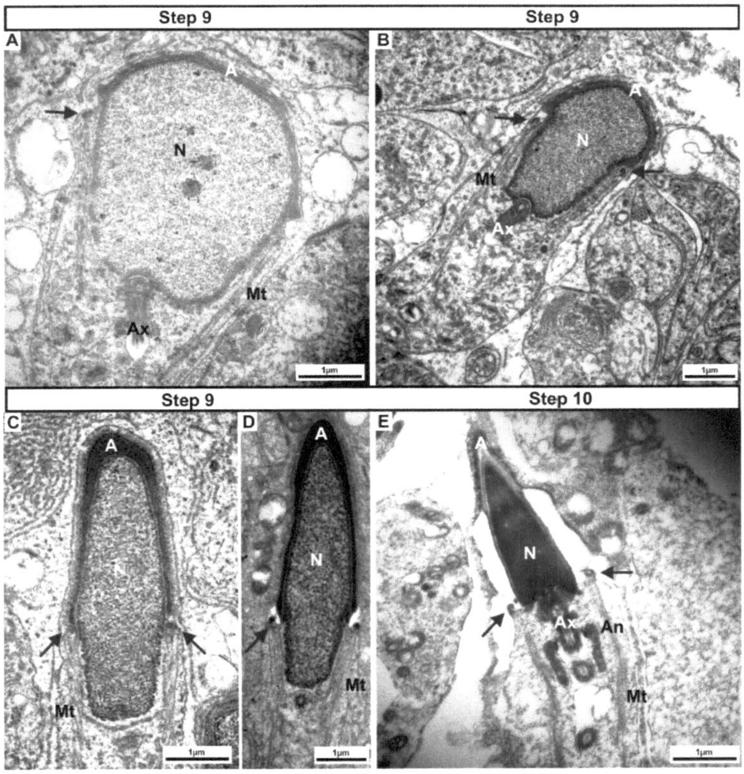

Fig. (7). Electron micrographs of the spermatids of *Platyrrhinus lineatus* (**A-C**), *Carollia perspicillata* (**D**), and *Eptesicus furinalis* (**E**) showing the spermiogenesis process (Acrosome phase). **A–D.** Early (**A**), middle (**B**), and late (**C-D**) step 9 spermatids. Note that major changes in nuclear elongation and condensation gradually occur in this step, when the longitudinal microtubule manchette (Mt), connected to the nuclear ring (arrows), has its fibers gradually shortened, pulling the acrosomes, making it cover the entire rostral end of the nucleus, which will assume an arrow shape (**E**). **E.** Step 10 spermatid. Note that at this step, the nuclear ring completes its dislocation to the posterior region of the nucleus, the annulus commences its caudal migration, and the organization of the midpiece begins. (A, acrosome; An, annulus; Ax, axoneme; Mt, microtubule manchette; N, nucleus).

Acrosome Phase

In the acrosome phase, the acrosome develops, covering the entire nucleus, and as a consequence, nuclear elongation and condensation take place (Fig. **7**). Nuclear elongation occurs when the fibers of the longitudinal microtubule manchette attach to the nuclear ring (Fig. **7A**) and gradually shorten, pulling the acrosomes (Figs. **7B**-**D**) and making them cover the entire rostral end of the nucleus, which now assumes an arrow shape (Fig. **7E**). Nuclear condensation occurs gradually and concomitantly with nuclear elongation. The flagellum develops considerably during this phase, with formation of the nuclear fossa, positioning of the centriolar adjunct, and elongation of the axoneme (Fig. **7E**).

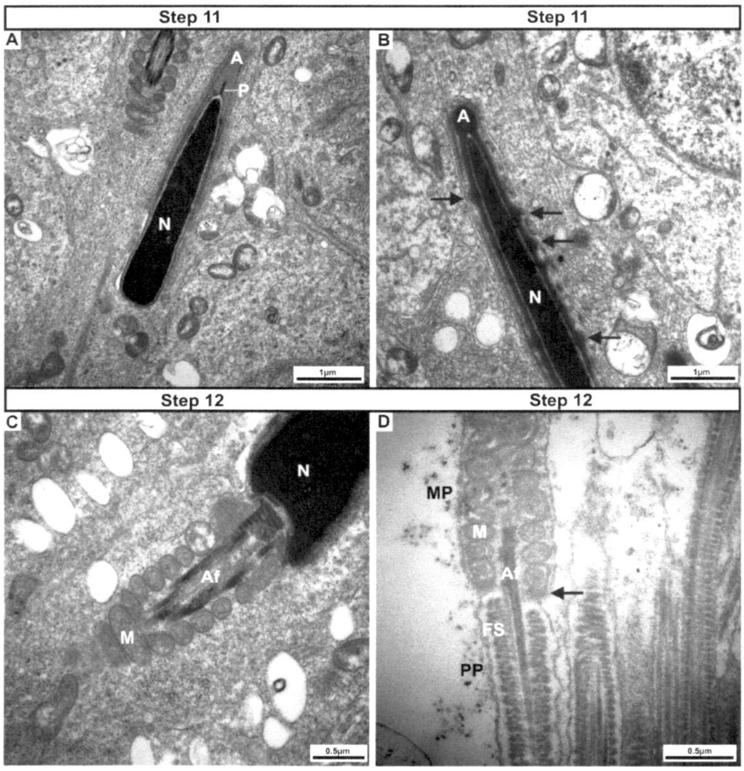

Fig. (8). Electron micrographs of the spermatids of *Molossus molossus* (**A**) and *Platyrrhinus lineatus* (**B-D**) showing the spermiogenesis process (Maturation phase). **A–B.** Step 11 spermatids. Note the beginning of the formation of the perforatorium (if it is present in the species) in the rostral region of the nucleus (**A**) and the development of secondary structures, like the acrosomal projections, forming lateral blisters on the spermatozoon surface (**B**, arrows). **C-D.** Step 12 spermatid. Note that at this step, the spermatids have achieved the basic format of the spermatozoa, with the compact and arrow-shaped head and the mid (**C**), principal (**D**), and end pieces already organized. (A, acrosome; Af, axial filament; FS, fibrous sheath; M, mitochondria; MP, midpiece; N, nucleus; P, perforatorium; PP, principal piece).

Maturation Phase

The maturation phase ends the differentiation process for the formation of spermatozoa. In this phase, the acrosome and nucleus complete their elongation and condensation, and the spermatozoon head assumes its final form (Fig. **8A**). Similarly, the mid, principal, and end pieces are organized, forming the functional flagellum (Figs. **8C-D**). All secondary structures (like acrosomal ornaments, perforatorium, *etc.*) seem to be built in this phase (Figs. **8A-B**) and, in the final step of this phase, the spermatozoa were release into the seminiferous lumen (spermiation process).

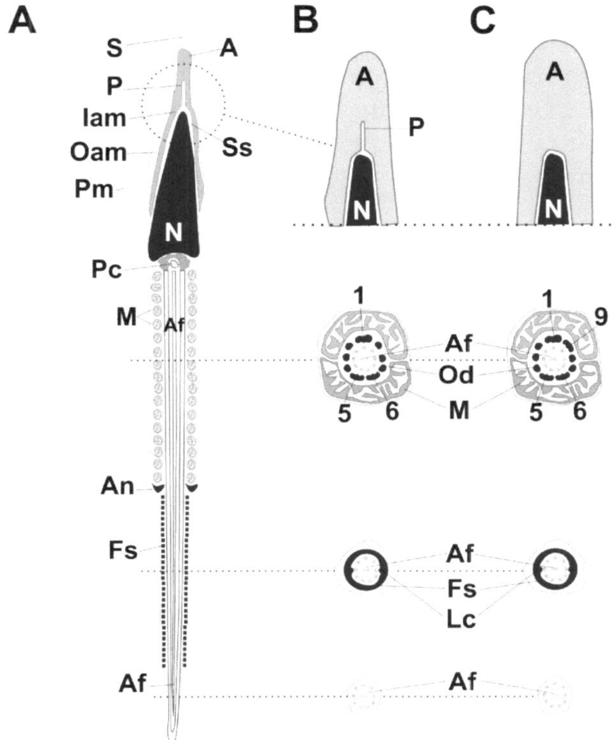

Fig. (9). Schematic representation of the ultrastructure of the spermatozoon of bats. **A.** Diagram of the basic bat spermatozoon in the longitudinal section. **B.** Diagram showing the presence of the perforatorium and the larger outer dense fibers 1, 5, and 6 in some species. **C.** Diagram showing the absence of the perforatorium and the larger outer dense fibers 1, 5, 6, and 9 in other species. (A, acrosome; Af, axial filament; An, annulus; Fs, fibrous sheath; Iam, inner acrosomal membrane; Lc, longitudinal column; M, mitochondria; N, nucleus; Oam, outer acrosomal membrane; Od, outer dense fibers; P, perforatorium; Pc, proximal centriole; Pm, plasma membrane; S, space between the plasma membrane and the outer acrosomal membrane; Ss, subacrosomal space).

ULTRASTRUCTURE OF BAT SPERMATOZOA

The general ultrastructure of the spermatozoon of bats seems to be relatively similar in most species already studied [2, 12, 19 - 21, 30, 31] and resembles the typical mammalian sperm model (Fig. **9A**), which consists of a dorsoventrally flattened, ovate, ensiform, or falciform head [32]. However, its morphology differs from other mammals such as marsupials [33, 34] and rodents [28] on account of a simpler spermatozoon head morphology, which indicates a pattern that is more closely related to the spermatozoa of humans and other primates [32, 35, 36].

The typical bat sperm has a dorsoventrally flattened, arrow-like head (Fig. **8**). The nucleus is arrow-shaped and occupies almost the entire head, presenting an extremely electron-dense and homogeneous chromatin. The acrosome has a highly homogeneous electron-dense matrix that extends up from the rostral region of the nucleus and covers it; it is externally surrounded by the outer acrosomal membrane and internally by the inner acrosome membrane, which is in contact with a reduced subacrosomal space (Fig. **8A**).

The base of the nucleus has a concavity where the flagellum attaches to the head (Fig. **8A**). The flagellum can be divided in three distinct portions: midpiece, principal piece, and end piece.

The initial portion of the flagellum corresponds to the midpiece. The midpiece is composed of the axial filament, which is surrounded by the nine outer dense fibers, and the mitochondrial sheath surrounds both. This sheath can be slightly shorter, containing approximately 30 to 40 mitochondrial gyres in some species [2, 12, 19 - 21], or longer, containing up to 160 in others [30]. In cross-section, the outer dense fibers show bilateral symmetry and a horseshoe fashion arrangement: fibers 3 and 8 remain aligned with the central fibrils of the axial filament complex, forming the dorsoventral axis (X-Y). This asymmetrically separates the complex, leaving one side with three fibers (1, 2, and 9) and the other with four (4, 5, 6, and 7). An annulus is found caudal to the last mitochondrial gyre. Immediately caudal to the annulus, the tail diameter lessens abruptly and gives rise to the principal piece (Fig. **8A**).

The principal piece is longer than the middle piece. In the upper region of the principal piece, projections from the fibrous sheath extend into the position of the outer dense fibers 3 and 8, forming the longitudinal column that supports the helical fibers of the fibrous sheath (Fig. **8A**). In addition, the end piece has only the axial filament enclosed by the plasma membrane (Fig. **8A**).

Although the majority of bat species has this typical pattern, some species-specific variations are observed. In some species, at the apical edge of the nucleus, the inner acrosomal membrane extends to the anterior region, to the middle of the acrosome, forming the perforatorium, which has a spear- or trident-like shape (Fig. **8B**). The perforatorium is absent or poorly developed in the Vespertilionidae

[19, 21, 25, 31, 37 - 39], little developed in the Rhinolophidae [24, 40], and well developed in the Molossidae [2, 41] and Phyllostomidae families [12, 20].

The morphology of the outer dense fibers present two different patterns: one with the outer dense fibers 1, 5, 6, and 9 larger than the others (Fig. **8C**) and the other pattern with only 1, 5, and 6 being larger (Fig. **8B**). The first pattern includes species of the families Pteropodidae [30], Rhinolophidae [24, 40, 42], Miniopteridae [42], Vespertilionidae [19, 21, 25, 31, 37], and some species of Phyllostomidae [20]. The second includes the Neotropical species *Molossus molossus* (Molossidae) [2] and *Platyrrhinus lineatus* (Phyllostomidae) [12].

Unusual characters in the spermatozoon head are also typical in some species of bats, such as the extraordinarily large, flat, and eccentrically inserted head with accordion-like folds of the Noctilionidae [10] and the many wavy acrosomal projections of the Molossidae family [2, 41].

ULTRASTRUCTURE OF THE SUPPORTING CELLS OF BATS

As in other mammals, the spermatogenesis process is assisted/regulated by two types of somatic cells: Sertoli and Leydig cells.

Sertoli Cells

The epithelium of the seminiferous tubules is mainly composed of the cytoplasm of Sertoli cells, which are associated with the basal lamina (Fig. **10A**) and exhibit extensive cytoplasmic processes that envelope all associated germ cells (Fig. **10B**) and proceed into the lumen. This cell type presents two different regions: basal and apical.

The basal region corresponds to the lowest part of the epithelium, which is closely associated with the basal lamina. The nucleus, which is large and irregularly shaped, is localized in this region. It presents a highly euchromatic and homogeneous nucleoplasm and a prominent nucleolus. The basal cytoplasm exhibits a high concentration of subcellular structures, including the endoplasmic reticulum, Golgi complex, mitochondria, and ribosomes (Fig. **10A**).

The apical region presents large compacted masses of smooth endoplasmic reticulum and, together with mitochondria, surrounds the differentiating spermatids and spermatocytes (Fig. **10B**).

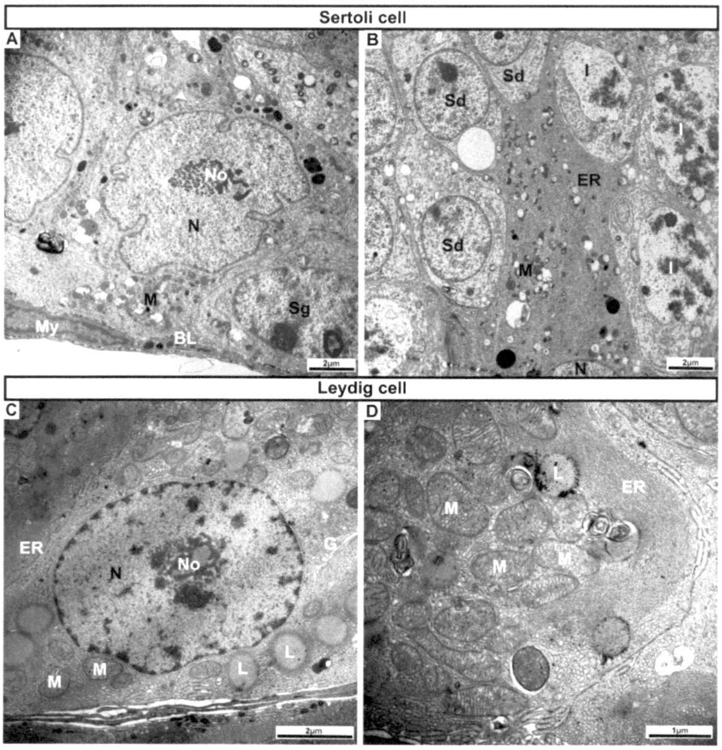

Fig. (10). Electron micrographs of the auxiliary cells of the bat *Molossus molossus*. **A–B.** Sertoli cells. **C-D.** Leydig cells. **A.** The basal region of a Sertoli cell. Note the large, irregularly elongated shaped nucleus with large infoldings and a prominent nucleolus as well as the perinuclear cytoplasm, which contains many mitochondrial clusters. **B.** The apical region of a Sertoli cell. Note the high concentration of subcellular structures, including endoplasmic reticulum and mitochondria. **C.** Note the large nuclei of the Leydig cell, with a greater portion of granular euchromatin and the single and large nucleolus, and the large amount of endoplasmic reticulum, mitochondria, and lipid droplets. **D.** Cytoplasm of Leydig cells. Note the large amount of endoplasmic reticulum, predominantly in smooth form, mitochondria, and lipid droplets. (BL, basal lamina; ER, endoplasmic reticulum; I, spermatocytes; L, lipid droplets; M, mitochondria; My, myoid cell; N, nucleus; No, nucleolus; Sp, spermatid).

Leydig Cells

The seminiferous tubules are surrounded by interstitial tissue composed of a great amount of extracellular matrix, nerve bundles, blood and lymphatic vessels, immune cells, and a great amount of Leydig cells.

Leydig cells are irregular polygonal cells found singly or in clusters within the interstitial tissue. Their nuclei are generally large, undefined (circular to oval), and possess a greater portion of granular euchromatin and a single large nucleolus (Fig. **10C**). The Leydig cell cytoplasm is vast and presents a highly secretory aspect, with a great portion of such cytoplasm occupied by the endoplasmic reticulum, predominantly in the smooth form, lipid droplets, and mitochondria (Fig. **10D**).

PERSPECTIVES

Despite the great number of species and the worldwide distribution, there are many gaps in our knowledge of the male reproduction of Chiroptera, with basic processes, such as spermatogenesis, described only in a few taxa. This scarce literature shows significant variations in morphology, physiology and endocrine control between different species of bats. Furthermore, many phylogenetic and evolutionary incongruences were observed in the Chiroptera taxa, with the relationships within the order still poorly understood. Thus, the analysis and comparison of the spermatogenesis process of different species of bats, in order to find morphological characteristics that demonstrate inter-specific, inter-generic, subfamilial or familial variation, may raise important morphological data that may contribute to the understanding of reproductive processes, as well as, may help in understanding the evolutionary and phylogenetic relationships of bats.

CONFLICT OF INTEREST

This research was supported by Grants 2009/16181–9 and 2009/03470–2 from the São Paulo State Research Foundation (FAPESP) and by Brazilian National Research and Development Council (CNPq) Processes: 300163/2008-8 and 301596/2011-5 research fellowships to SRT and Process: 302008/2010-1 to EMV. The authors disclose no conflicts of interest.

ACKNOWLEDGEMENTS

Mateus R. Beguelini contributes to all data analysis/interpretation and to the draft version and development of the chapter; Sebastião R. Taboga and Eliana Morielle-Versute critically revised and approved the chapter. The authors thank Luiz Roberto Falleiros Junior for technical help.

REFERENCES

[1] Matsumoto AM. Spermatogenesis. In: Adashi EY, Rock JA, Rosenwaks Z, Eds. Reproductive Endocrinology, Surgery and Technology. Philadelphia: Lippincott-Raven 1996; pp. 360-84.

[2] Beguelini MR, Taboga SR, Morielle-Versute E. Ultrastructural characteristics of spermatogenesis in Pallas's mastiff bat, Molossus molossus (Chiroptera: Molossidae). Microsc Res Tech 2012; 75(7): 856-68.
[http://dx.doi.org/10.1002/jemt.22005] [PMID: 22253210]

[3] Costa GC, Vieira GH, Teixeira RD, Garda AA, Colli GR, Báo SN. An ultrastructural comparative study of the sperm of *Hyla pseudopseudis, Scinax rostratus* and *S. squalirostris* (Amphibia: Anura: Hylidae). Zoomorphology 2004; 123: 191-7.
[http://dx.doi.org/10.1007/s00435-004-0101-z]

[4] Garda AA, Costa GC, Colli GR, Báo SN. Spermatozoa of Pseudinae (Amphibia, Anura, Hylidae), with a test of the hypothesis that sperm ultrastructure correlates with reproductive modes in anurans. J Morphol 2004; 261(2): 196-205. [http://dx.doi.org/10.1002/jmor.10238] [PMID: 15216524]

[5] Vieira GH, Colli GR, Báo SN. Phylogenetic relationships of corytophanid lizards (Iguania, Squamata, Reptilia) based on partitioned and total evidence analyses of sperm morphology, gross morphology, and DNA data. Zool Scr 2005; 34: 605-25. [http://dx.doi.org/10.1111/j.1463-6409.2005.00208.x]

[6] Jeong SJ, Park JC, Kim HJ, *et al.* Comparative fine structure of the epididymal spermatozoa from three Korean shrews with considerations on their phylogenetic relationships. Biocell 2006; 30(2): 279-86. [PMID: 16972552]

[7] Clermont Y. Kinetics of spermatogenesis in mammals: seminiferous epithelium cycle and spermatogonial renewal. Physiol Rev 1972; 52(1): 198-236. [PMID: 4621362]

[8] Russell LD, Ettlin RA, Sinha-Hikin AP, Clegg ED. Mammalian spermatogenesis. In: Russell LD, Ettlin RA, Sinha Hikin AP, Clegg ED, Eds. Histological and Histopathological Evaluation of the Testis. Clearwater: Cache River Press 1990; pp. 1-40.

[9] Lin M, Harman A, Rodger JC. Spermiogenesis and spermiation in a marsupial, the tammar wallaby (*Macropus eugenii*). J Anat 1997; 190(Pt 3): 377-95. [http://dx.doi.org/10.1046/j.1469-7580.1997.19030377.x] [PMID: 9147224]

[10] Phillips DM, Rasweiler JJ IV, Muradali F. Giant, accordioned sperm acrosomes of the greater bulldog bat, *Noctilio leporinus.* Mol Reprod Dev 1997; 48(1): 90-4. [http://dx.doi.org/10.1002/(SICI)1098-2795(199709)48:1<90::AID-MRD11>3.0.CO;2-#] [PMID: 9266765]

[11] Lin M, Jones RC. Spermiogenesis and spermiation in a monotreme mammal, the platypus, *Ornithorhynchus anatinus.* J Anat 2000; 196(Pt 2): 217-32. [http://dx.doi.org/10.1046/j.1469-7580.2000.19620217.x] [PMID: 10739018]

[12] Beguelini MR, Puga CC, Taboga SR, Morielle-Versute E. Ultrastructure of spermatogenesis in the white-lined broad-nosed bat, *Platyrrhinus lineatus* (Chiroptera: Phyllostomidae). Micron 2011; 42(6): 586-99. [http://dx.doi.org/10.1016/j.micron.2011.02.004] [PMID: 21458280]

[13] Jamieson BG. Evolution of tetrapod spermatozoa with particular reference to amniotes. In: Jamieson BG, Ausio J, Justine JL, Eds. Advances in spermatozoal phylogeny and taxonomy. Paris: Mémorielles del Muséum National d'Histoire Naturelle 1995; pp. 343-58.

[14] Singwi MS, Lall SB. Spermatogenesis in the non-scrotal bat-*Rhinopoma kinneari* Wroughton (microchiroptera: mammalia). Acta Anat (Basel) 1983; 116(2): 136-45. [http://dx.doi.org/10.1159/000145735] [PMID: 6880591]

[15] McGuckin MA, Blackshaw AW. Cycle of the seminiferous epithelium in the grey-headed fruit bat, *Pteropus poliocephalus.* Aust J Biol Sci 1987; 40: 203-10. [http://dx.doi.org/10.1071/BI9870203]

[16] Saidapur SK, Patil SB. Kinetics of spermatogenesis in megachiropteran bat, *Rousettus leschenaulti* (Desmarset): seminiferous epithelial cycle, frequency of stages, spermatogonial renewal and germ cell degeneration. Indian J Exp Biol 1992; 30(11): 1037-44. [PMID: 1293027]

[17] Morais DB, de Paula TA, Barros MS, Balarini MK, de Freitas MB, da Matta SL. Stages and duration of the seminiferous epithelium cycle in the bat *Sturnira lilium.* J Anat 2013; 222(3): 372-9. [http://dx.doi.org/10.1111/joa.12016] [PMID: 23305159]

[18] Beguelini MR, Moreira PR, Faria KC, Marchesin SR, Morielle-Versute E. Morphological characterization of the testicular cells and seminiferous epithelium cycle in six species of Neotropical bats. J Morphol 2009; 270(8): 943-53.
[http://dx.doi.org/10.1002/jmor.10731] [PMID: 19248152]

[19] Beguelini MR, Taboga SR, Morielle-Versute E. Ultrastructural characteristics of the spermatogenesis during the four phases of the annual reproductive cycle of the black *myotis* bat, *Myotis nigricans* (Chiroptera: Vespertilionidae). Microsc Res Tech 2013; 76(10): 1035-49.
[http://dx.doi.org/10.1002/jemt.22264] [PMID: 23857678]

[20] Beguelini MR, Bueno LM, Caun DL, Taboga SR, Morielle-Versute E. Ultrastructure of spermatogenesis in the short-tailed fruit bat, *Carollia perspicillata* (Chiroptera: Phyllostomidae: Carollinae). J Morphol 2014; 275(1): 111-23.
[http://dx.doi.org/10.1002/jmor.20202] [PMID: 24142890]

[21] Bueno LM, Beguelini MR, Comelis MT, Taboga SR, Morielle-Versute E. Ultrastructure of spermatogenesis, spermatozoon and processes of testicular regression and recrudescence in *Eptesicus furinalis* (Chiroptera: Vespertilionidae). Anim Reprod Sci 2014; 148(3-4): 228-44.
[http://dx.doi.org/10.1016/j.anireprosci.2014.05.018] [PMID: 24954586]

[22] Petersen C, Söder O. The sertoli cell--a hormonal target and 'super' nurse for germ cells that determines testicular size. Horm Res 2006; 66(4): 153-61.
[PMID: 16804315]

[23] Lodish H, Berk A, Matsudaira P, Eds. Molecular Cell Biology. New York: WH Freeman & Co 2005.

[24] Lee JH, Choi BJ, Son SW. Spermiogenesis in the Korean Greater Horseshoe Bat, *Rhinolophus ferrumequinum korai.* Kor J Electron Microsc 1992; 22: 97-117.

[25] Lee JH. Cell differentiation and ultrastructure of the seminiferous epithelium in *Myotis macrodactylus.* Kor J Electron Microsc 2003; 33(1): 25-39.

[26] Morigaki T, Kurohmaru M, Kanai Y, *et al.* Cycle of the seminiferous epithelium in the Java fruit bat (*Pteropus vampyrus*) and the Japanese lesser horseshoe bat (*Rhinolophus cornutus*). J Vet Med Sci 2001; 63(7): 773-9.
[http://dx.doi.org/10.1292/jvms.63.773] [PMID: 11503905]

[27] Kurohmaru M, Kobayashi H, Hattori S, Nishida T, Hayashi Y. Spermatogenesis and ultrastructure of a peculiar acrosomal formation in the musk shrew, *Suncus murinus.* J Anat 1994; 185(Pt 3): 503-9.
[PMID: 7649786]

[28] Segatelli TM, Almedia CC, Pinheiro PF, Martinez M, Padovani CR, Martinez FE. Ultrastructural study of acrosome formation in mongolian gerbil (*Meriones unguiculatus*). Tissue Cell 2000; 32(6): 508-17.
[http://dx.doi.org/10.1016/S0040-8166(00)80007-5] [PMID: 11197233]

[29] Suphamungmee W, Wanichanon C, Vanichviriyakit R, Sobhon P. Spermiogenesis and chromatin condensation in the common tree shrew, *Tupaia glis.* Cell Tissue Res 2008; 331(3): 687-99.
[http://dx.doi.org/10.1007/s00441-007-0557-5] [PMID: 18095001]

[30] Crichton EG, Krutzsch PH, Eds. Reproductive Biology of Bats. London: Academic Press 2000.

[31] Fawcett DW, Ito S. The fine structure of bat spermatozoa. Am J Anat 1965; 116: 567-609.
[http://dx.doi.org/10.1002/aja.1001160306] [PMID: 14324688]

[32] Fawcett DW. A comparative view of sperm ultrastructure. Biol Reprod Suppl 1970; 2: 90-127.
[http://dx.doi.org/10.1095/biolreprod2.Supplement_2.90] [PMID: 12254595]

[33] Johnston SD, Daddow L, Carrick FN, Jamieson B. Observations of spermiogenesis and epididymal sperm maturation in the rufous hare wallaby, *Lagorchestes hirsutus* (Metatheria, Mammalia). Acta Zool 2004; 85: 53-8.
[http://dx.doi.org/10.1111/j.0001-7272.2004.00157.x]

[34] Lloyd S, Carrick F, Hall L. Unique features of spermiogenesis in the Musky Rat-kangaroo: reflection of a basal lineage or a distinct fertilization process? J Anat 2008; 212(3): 257-74. [http://dx.doi.org/10.1111/j.1469-7580.2008.00861.x] [PMID: 18304206]

[35] Hoffer AP, Shalev M, Frisch DH. Ultrastructure and maturational changes in spermatozoa in the epididymis of the pigtailed monkey, *Macaca nemestrina.* J Androl 1981; 3: 140-6. [http://dx.doi.org/10.1002/j.1939-4640.1981.tb00608.x]

[36] Holt WV, Moore HD. Ultrastructural aspects of spermatogenesis in the common marmoset (*Callithrix jacchus*). J Anat 1984; 138(Pt 1): 175-88. [PMID: 6423594]

[37] Son SW, Lee JH, Cheon HM. Spermiogenesis in the Korean Daubenton's Bat (*Myotis daubentonii ussuriensis*). Dev Reprod 1997; 1: 9-24.

[38] Kim HH, Lee JH. Fine Structure of the Sperm in the *Myotis daubentonii ussuriensis.* Kor J Micros 2011; 41(1): 31-5.

[39] Lee YR, Lee JH. Sperm Ultrastructure of *Pipistrellus savii velox.* Kor J Micros 2011; 41(2): 117-22.

[40] Oh YK, Mori T, Uchida TA. Spermiogenesis in the Japanese Greater Horseshoe Bat, *Rhinolophus ferrumequinum* Nippon. J Fac Agric Kyushu Univ 1985; 29: 203-9.

[41] Breed WG, Leigh C. Sperm head morphology of Australian molossid bats with special reference to the acrosomal structure. Mammalia 1985; 49: 403-6. [http://dx.doi.org/10.1515/mamm.1985.49.3.403]

[42] Sang-Sick K, Jung-Hun L, Sung-Won S, Byung-Jin C. Morphological comparison of spermatozoa in the Korean greater horseshoe bat (Rh*inolophus ferrumequinum korai*) and long-fingered bat (*Miniopterus schreibersi fuliginosus*). Kor J Electron Microsc 1999; 29: 1-10.

CHAPTER 2

Male Reproductive Hormonal Regulation in Bats and the Variations Associated with Testicular Regression

Mateus Rodrigues Beguelini[1,*], **Sebastião Roberto Taboga**[2] and **Eliana Morielle-Versute**[3]

[1] *Center of Biological and Health Sciences, UFOB – Universidade Federal do Oeste da Bahia, Barreiras, Bahia, 47808-021, Brazil*

[2] *Department of Biology, UNESP – Univ. Estadual Paulista, São José do Rio Preto, São Paulo, 15054-000, Brazil*

[3] *Department of Zoology and Botany, UNESP – Univ. Estadual Paulista, São José do Rio Preto, São Paulo,15054-000, Brazil*

Abstract: The reproductive process requires the coordinated action of a large variety of peptides and steroid hormones, each one playing an important role in the normal functioning of the reproductive organs. In male bats, this control is linked to the correct functioning of the hypothalamic-pituitary-testicular axis, combined to the paracrine communication between Sertoli, Leydig and germ cells. The hypothalamus produces and secretes gonadotropin-releasing hormone (GnRH), which acts on the pituitary gland stimulating the production of follicle-stimulating hormone (FSH) and luteinizing hormone (LH). Both hormones act directly on the testicular somatic cells in order to stimulate and regulate their function and to control the spermatogenesis process. Testicular steroids (androgens and estrogens) in turn stimulate the seminiferous tubules and all secondary sex structures (epididymis and reproductive accessory glands, mainly the prostate), as well as provide feedback to the hypothalamus and pituitary. The hormonal regulation of bats differs interspecifically, with some species presenting the typical mammalian pattern; others showing unique variations, possibly imposed by the hibernation period; until species that have developed different adaptations, but without the imposition of a period of hibernation. One of these adaptations is the testicular regression, which is a deactivation process of the testicular tissue that is triggered by changes in the pattern of hormonal regulation. Thus, in this chapter, we seek to discuss different patterns of hormonal regulation of bats and some aspects of testicular regression.

Keywords: Androgen Receptor, Bat, Chiroptera, Estradiol, Estrogen, Hormonal Regulation, Reproduction, Spermatogenesis, Testicular Regression, Testis, Testosterone.

* **Corresponding author Mateus Rodrigues Beguelini:** Rua Prof. José Seabra de Lemos nº 316, Recanto dos Pássaros, 47808-021, Barreiras, Bahia, Brazil; Tel: +55 77 36143215; E-mail: mateus.beguelini@ufob.edu.br

Edith Arenas-Ríos (Ed.)

MAMMALIAN HORMONAL REGULATION

The reproductive process requires the coordinated action of a large variety of peptides and steroid hormones, each one playing an important role in the normal functioning of the reproductive organs. These hormonal messengers are essential not only in the coordination of the development and maturation of the organs in juveniles, but also in the regulation and maintenance of the normal function of different cell types in adults [1].

The Hypothalamic-Pituitary-Testicular Axis (HPT)

In male bats, as well as in other mammals, the control of reproduction is linked to the correct functioning of the hypothalamic-pituitary-testicular axis (HPT), combined to the paracrine communication between Sertoli, Leydig and germ cells [1 - 3]. The cascade of reactions begins with the production and secretion of gonadotropin-releasing hormone (GnRH) by the hypothalamus, which acts on the pituitary gland stimulating the production of gonadotropins: follicle-stimulating hormone (FSH) and luteinizing hormone (LH). Both hormones act directly on the testicular somatic cells in order to stimulate and regulate their function and to control the spermatogenesis process. Testicular steroids (androgens and estrogens) in turn stimulate the seminiferous tubules and all secondary sex structures, and provide feedback to the hypothalamus and pituitary (Fig. **1**) [2, 3].

The main testicular target of the FSH seems to be the Sertoli cells (Fig. **1**), since they are the only testicular cells expressing its receptor [4]. Genetic and pharmacological studies suggest that the primary role of FSH in spermatogenesis is to stimulate the proliferation of Sertoli cells during the pre-pubertal period of the development. There is a broad consensus that the total number of Sertoli cells largely determines the number of associated germ cells and the final sperm production of the adult [5]. However, its role in the regulation of the spermatogenesis in adults has been extensively evaluated and the studies demonstrate that FSH also plays an important role in the regulation of spermatogonial differentiation, although the detailed mechanism of its action is still uncertain [6].

Similarly, LH seems to have the testicular Leydig cells as its main target (Fig. **1**), once these are the only testicular cells expressing its receptor (LHR). LH binds to its receptor and stimulates Leydig cells to produce androgens, including testosterone (T), 5α-dihydrotestosterone (DHT) and androstenedione (AND). The regulation of T synthesis by Leydig cells appears to be the only essential function of LH in adults, since LHR knockout mice, treated with exogenous T, had normal spermatogenesis [7].

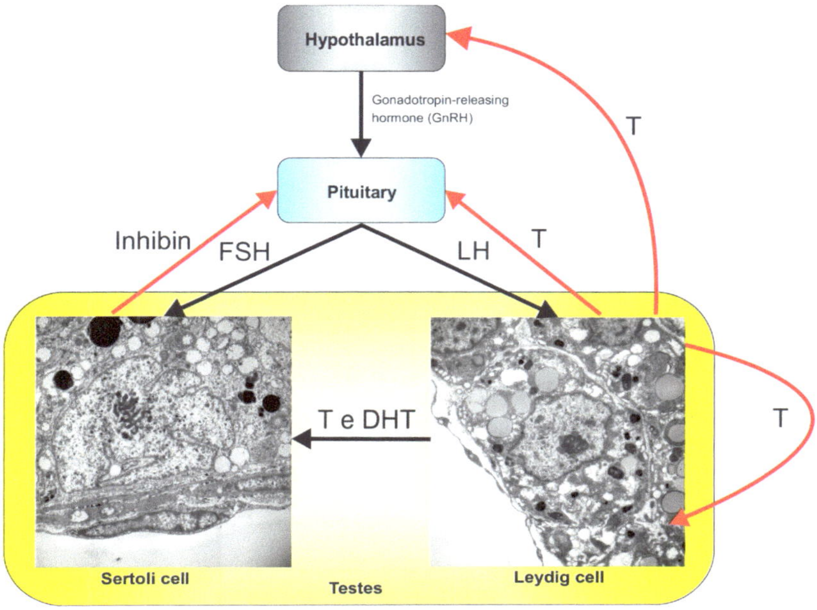

Fig. (1). Schematic representation showing the hormonal regulation of the hypothalamic-pituitary-testicular axis (HPT). Black arrows indicate stimulants effects while red arrows show feedback effects. (DHT, dihydrotestosterone; FSH, follicle-stimulating hormone; LH, luteinizing hormone; T, testosterone). Electron micrographs of the bat *Platyrrhinus lineatus*.

The Paracrine Regulation of the Testes

The main role of testicular androgens, especially T and DHT, is connected to the initiation, maintenance and restoration of spermatogenesis [6]. Although details of the cellular and molecular mechanisms of action of these hormones still remain poorly defined, there is a broad consensus that their biological actions within the testes are primarily mediated *via* androgen receptor (AR), which is located mainly in the Sertoli (Fig. **2A**) and Leydig cells (Fig. **2B**) [1, 8, 9]. AR expression in Leydig cells is connected to an autocrine feedback on the regulation of T production and conversion of their metabolites (Fig. **1**). Conversely, in Sertoli cells, the AR expression and activation trigger their nuclear translocation and activate their regulatory function of the transcription, causing profound effects on spermatogenesis. The AR activation (T-AR bidding) is required in Sertoli cells in at least three stages of the spermatogenesis process: in the progression through meiosis I; in the elongation process of spermatids; and finally, during the final stages of spermiogenesis [1].

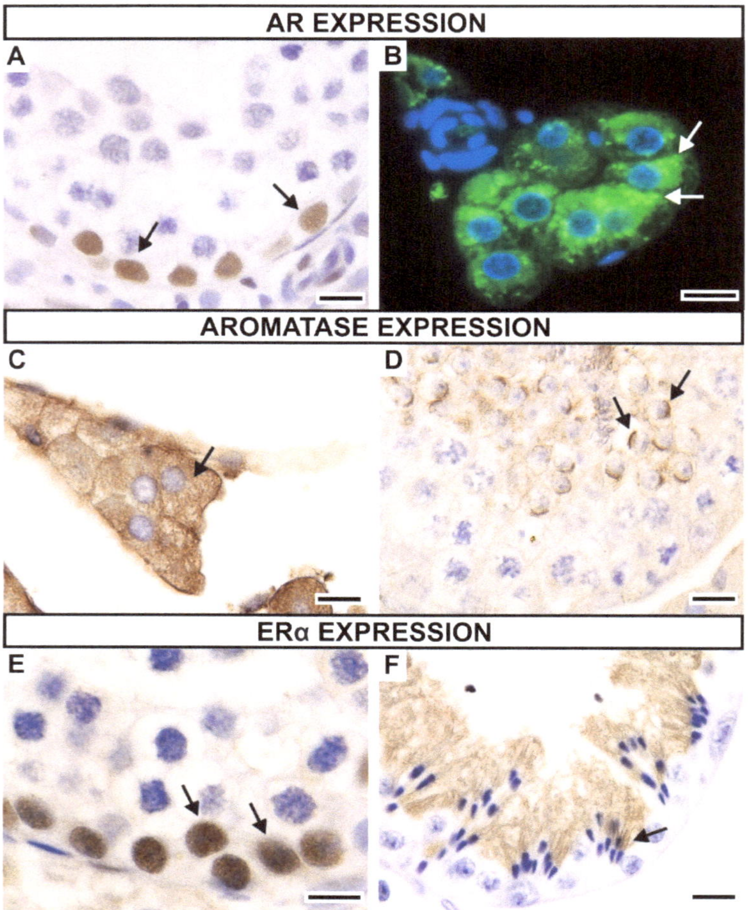

Fig. (2). Testicular sections showing the expression pattern of AR (**A-B**), aromatase (**C-D**) and ERα (**E-F**) in the testes of the bat *Myotis nigricans* (Vespertilionidae: Chiroptera), after specific immunohistochemical reactions. Figures modified from Beguelini *et al.*, 2014 [9, 12]. **A.** Note the nuclear expression of the AR in Sertoli cells (arrows). **B.** Observe the green fluorescence showing the expression of the AR in Leydig cells (arrows). **C.** Note the cytoplasmic expression of aromatase in Leydig cells (arrow). **D.** Observe the expression of aromatase in spermatids (arrows). **E.** Note the expression of ERα in Sertoli cells (arrows). **F.** Observe the expression of ERα in spermatids (arrows). Scale bars = 10μm.

On the other hand, these are not the only hormones that act on the regulation of the testicular function. The involvement of estrogens in the regulation of testicular physiology, including spermatogenesis, has been proven by many studies [10, 11]. Estrogens are locally produced by several types of cells in the testis, including Leydig (Fig. **2C**) and Sertoli cells and, in adults, particularly germ cells (Fig. **2D**) [12]. They act by binding and activating specific receptors, particularly the estrogen receptors alpha (ERα) and beta (ERβ), which are expressed in several testicular cells, such as Sertoli cells, spermatogonia, elongated spermatids and spermatozoa (Figs. **2E-F**) [12]. This expression pattern suggests a paracrine action on spermatogenesis [13]; however, circulating estrogen may also have some impact on testicular function *via* endocrine mechanisms of feedback to the hypothalamus and pituitary [14, 15]. In addition, the study of ERα, ERβ and aromatase knockout mice, showed that males in this situation are infertile, indicating that the normal action of estrogen is essential for spermatogenesis [16, 17].

These studies showed that all reproductive hormones, FSH and LH (in an endocrine way) and androgens and estrogens (in an endocrine and paracrine way), act in a coordinated manner to activate and regulate the testicular function.

Regulation of Secondary Reproductive Organs

Similarly, since the second half of the twentieth century, several researchers have demonstrated that the main regulatory factor of the function of the secondary reproductive organs (epididymis and reproductive accessory glands – mainly the prostate) is the androgens synthesized by the testes. T is primarily identified as the main regulatory molecule; however, several studies have provided increasing evidence that estrogens also play specific regulatory roles in their morphophysiologies [18, 19].

Epididymal Regulation

Testicular T reaches the epididymis *via* two distinct routes: 1) *via* luminal transportation and 2) *via* circulation [20]. In both cases, the T is rapidly converted to DHT by the enzyme steroid 5α-reductase, which is mainly present in the epithelial cells of the initial segment and caput epididymis [21, 22]. This conversion is important, because DHT has an increased excitatory activity by having a greater affinity to AR [23]. The T may also be converted to estradiol (E2) by the P450 aromatase enzyme, which is expressed in the luminal spermatozoa and epithelial and interstitial epididymal cells (Fig. **3**) [18].

The binding of DHT to the AR of epididymal cells triggers different cascades of reactions that are required to maintain the epithelial cell morphology, to prevent cell death and to regulate their protein expression and secretion, which ensures the maturation and storage of the spermatozoa [24]. Moreover, the estrogens actively regulate fluid reabsorption and the expression of some proteins involved in sperm membrane remodeling and in the initiation of sperm motility [25, 26].

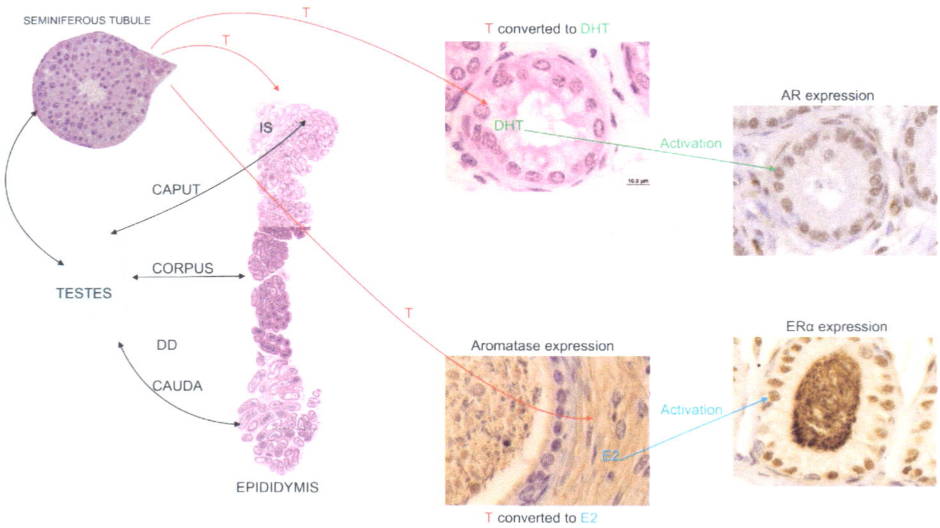

Fig. (3). Schematic representation showing the action of testicular testosterone in the endocrine regulation of the epididymis. The testicular T that reaches the initial segment and caput epididymis is rapidly converted to DHT by their epithelial cells (**A**). This conversion is important because, the binding of DHT to the AR of epididymal cells (**B**) triggers different cascades of reactions that are required to maintain the epithelial cell morphology, to prevent cell death and to regulate their protein expression and secretion, which ensures the maturation and storage of the spermatozoa. Conversely, the T may also be converted to E2 by the P450 aromatase enzyme, which is expressed in the luminal spermatozoa and epithelial and interstitial epididymal cells (**C**). This conversion is important because, the binding of E2 to the ERs of epididymal cells (**D**) actively regulate their fluid reabsorption and the expression of some proteins involved in sperm membrane remodeling and in the initiation of sperm motility, thus, directly influencing in the animal reproductive success. (DD, deferens duct; DHT, dihydrotestosterone; E2, estradiol; IS, initial segment; T, testosterone). Histologic sections of the bats *Molossus molossus* (**A**, caput epididymis with haematoxylin-eosin stain) and *Myotis nigricans* (**B**, AR expression in caput epididymis; **C**, Aromatase expression in cauda epididymis; **D**, ERα expression in corpus epididymis). Scale bars = 10μm. Figures modified from Beguelini *et al.*, 2014 [9, 12].

Prostatic Regulation

The mature prostate is mainly regulated by circulating T, which stimulates/regulates epithelial and stromal cells, and triggers different paracrine signaling pathways, to maintain their fully differentiated growth-quiescent epithelium [27, 28]. Stimulation *via* epithelial cells is given by the transformation of T into DHT, which is catalyzed by the 5α-reductase enzyme mainly present in the cytoplasm of epithelial cells [29]. The binding of DHT to the AR of epithelial cells triggers different cascades of reactions that primarily regulate their production and secretion. Similarly, the stimulation *via* stromal cells is provided by the conversion of T into E2, which is catalyzed by the aromatase enzyme, mainly present in the cytoplasm of stromal cells [19]. Estradiol is essential for normal tissue homeostasis within the prostate, and has opposing dual roles in prostatic tissue, with the activation of ERα leading to proliferation and the activation of estrogen ERβ mediating antiproliferative effects (apoptosis), which balance the proliferative action of androgens on the epithelia (Fig. **4**) [30, 31].

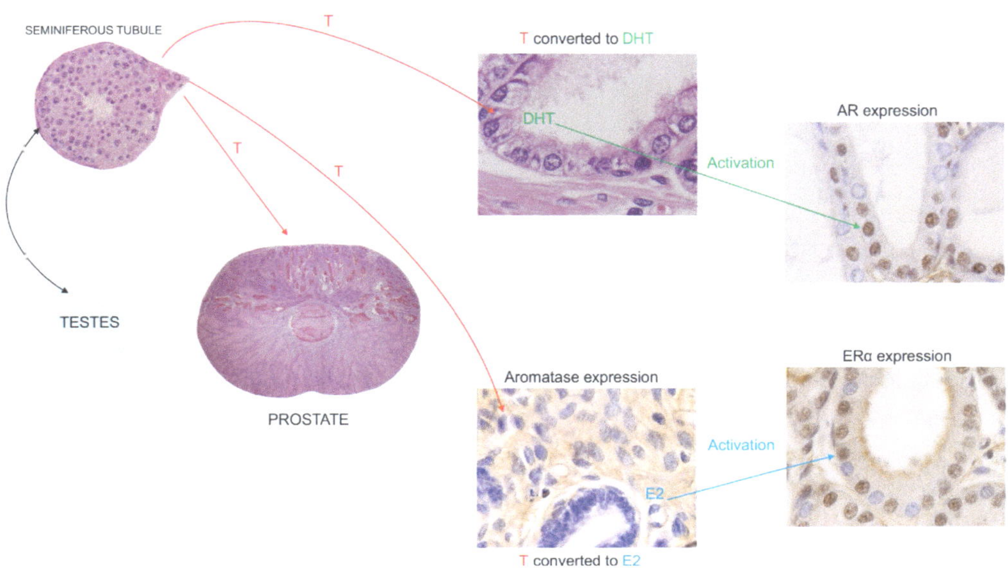

Fig. (4). Schematic representation showing the action of testicular testosterone in the endocrine regulation of the prostate. When the circulating T reaches the prostate, they may be converted into DHT by the epithelial cells (**A**). This conversion is important because, the binding of DHT to the AR (**B**) of epithelial cells directly regulate their production and secretion. Similarly, the stromal cells may convert T into E2 (**C**), and the binding of E2 to the ERs (**D**) is essential for the proper functioning of the prostate. (DHT, dihydrotestosterone; E2, estradiol; T, testosterone). Histologic sections of the bat *Myotis nigricans* (**A**, Dorsal prostate stained with haematoxylin-eosin; **B**, AR expression in dorsal prostate; **C**, Aromatase expression in dorsal prostate; **D**, ERα expression in dorsal prostate). Scale bars = 10μm. Figures modified from Beguelini *et al.*, 2014 [9, 12].

HORMONAL REGULATION IN BATS

Among mammals, bats exhibit the second largest number of species and the second largest territorial dispersion pattern [32]. This shows that the order Chiroptera has undergone a rich adaptive radiation, which has produced a high variation in reproductive patterns, structures and mechanisms, and triggers the development of many unique characteristics, such as: testicular regression; prolonged sperm storage in the cauda epididymis and/or in the uterine horn; asynchrony between spermatogenesis and copulation; delayed ovulation, fertilization and implantation, *etc.* [33]. Thus, not surprisingly, the hormonal regulation of bats is often very different from the mammalian pattern.

Despite the large number of species and reproductive adaptations, relatively few studies have investigated the hormonal control of reproduction in bats, with the existing data being limited to a few isolated measurements of the plasma levels of T or other hormones, and the location of the cells that produce gonadotropins [3, 33]. However, these studies show seasonal variations in the production of FSH, LH and T, as well as their receptors, in some bat species, with these oscillations generally linked to annual reproductive variations [3].

The literature shows that the hormonal regulation of bats differs in an interspecific way, with some species presenting the typical mammalian pattern; others showing unique adaptations, possibly imposed by the hibernation period; until species that have developed different adaptations, but without the imposition of a period of hibernation.

Many species of bats present the typical mammalian pattern of reproduction, presenting a continuously active process of spermatogenesis throughout the year, with the production of spermatozoa by the testes in all months [8, 34, 35]. Despite this constant production, peaks of spermatogenic production are frequently observed, mainly linked to breeding periods. Species with this pattern, such as the phyllostomid bat *Artibeus planirostris* [8], tend to have a more constant and low variable production of T and its receptors during the year, with peaks in production only in months that precede the breeding periods.

The constant annual expression of T stimulates the continuous production of spermatozoa by the testes and keeps partly active the secondary reproductive organs (epididymis and prostate), which provide, to many of these animals, the ability to breed any time of the year, depending mainly on the receptivity of females. In months preceding the breeding periods, the increase in production of T and its receptors increase the production of spermatozoa and, endocrinologically, trigger the hypertrophy of the epididymis and prostate, in order to produce and mature sufficient sperm for reproduction in a harem system [8]. Thus, in this pattern, the increase or decrease in the functionality of the epididymis and prostate occurs in direct synchrony with the testicular cycle, in order to prepare the animal for the breeding period.

Conversely, other species, such as the emballonurid bats *Taphozous georgianus* [34, 36] and *T. longimanus* [37], have an asynchrony in the activity of the testes and reproductive accessory glands (RAGs – prostate, seminal vesicle, *etc*.). In these animals, the hypertrophy of the RAGs has an inverse correlation with the testicular activity, with the maximum hypertrophy occurring in periods where the level of plasma T and sperm production has decreased.

The asynchrony between the activity of primary and secondary sexual organs has also been observed in almost all male bats that hibernate [33, 38]. However, this is not nearly the main reproductive adaptation that these animals have. The majority of temperate bats are seasonal breeders (monoestrous), with this characteristic being directly associated with hibernation and low temperatures [38]. The severe temperatures and the prolonged hibernation period during the winter restrict the reproductive process of hibernating animals to the summer, with spermatogenesis and copulation occurring during this period [33, 38]. As spermatogenesis occurs only in summer or into early autumn, after this period the seminiferous epithelium undergoes a process of deactivation, entering rest until its reactivation at the end of hibernation. During this period, the seminiferous tubules are characterized by a prepubertal-like morphology, in that they contain only spermatogonia and inactive Sertoli cells (Fig. **5**) [38].

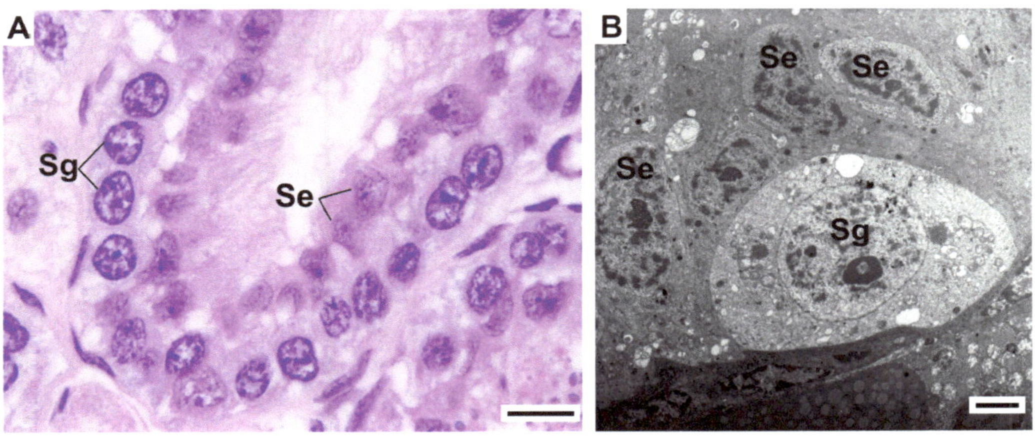

Fig. (5). Testicular sections of the bat *Myotis nigricans* showing the seminiferous epithelium composition during the regressed period, light microscopy (**A**) and Transmission Electron Microscopy (**B**). Note the prepubertal-like morphology, with the seminiferous epithelium containing only spermatogonia (Sg) and inactive Sertoli cells (Se). Scale bars: **A** = 10µm; **B** = 2µm.

This deactivation process, also called "testicular regression", is triggered by changes in the pattern of hormonal regulation.

Hormonal Regulation of Testicular Regression in Hibernating Bats

As already mentioned, few studies have investigated in detail the hormonal control of reproduction in bats. Even though we know that the hormonal regulation of reproduction of an organism is carried out by the interaction of various hormones, receptors and peptides produced by different cell types, studies on hibernating bats suggest that levels of LH are highest in periods when bats resume activity following their arousal from hibernation. This peak of LH restarts the Leydig cell activity and, consequently, the testicular production of T, which reactivates the Sertoli cells and triggers a period of intense mitotic activity, restarting the spermatogenesis process. In general, the plasma T levels tend to have their greatest value in this period, which leads to the reactivation of secondary reproductive organs. The intense spermatogenic production extends for a brief period, which varies slightly in duration according to species [33, 38].

In the months preceding hibernation, the LH levels reach their lowest value, which reduces the stimulus for T production in the Leydig cells and, consequently, leads to the deactivation of the spermatogenesis and the occurrence of testicular regression. The pattern of testicular deactivation is maintained during hibernation due to circulating LH levels not being sufficient to maintain a T production that supports the spermatogenesis process [33, 38].

Thus, hibernating bats tend to show an annual reproductive cycle composed of one peak of production of LH and associated hormones (T, DHT, E2, *etc.*), generally associated with the warm months (spring and summer), which stimulates the reactivation of the spermatogenesis and a large spermatic production, and a period of drastic reduction in circulating levels of these hormones, generally associated with the cold months (end of autumn and winter) and that causes the testicular regression.

Hormonal Regulation of Testicular Regression in Nonhibernating Bats

Although testicular regression presents persuasive features, which indicate that this process may have developed as an evolutionary adaptation to survival in face of harsh environments and hibernation, different studies have shown the existence of this process in the annual reproductive cycle of some species of tropical and subtropical nonhibernating bats [9, 12, 18, 19, 39 - 42]. The reason for the existence of testicular regression in species that are not subject to drastic environmental conditions and hibernation periods is still unknown; however, it is noted that the hormonal regulation patterns tend to be more complex and varied in these animals than those of hibernating animals.

Most species of nonhibernating bats that undergo the process of testicular regression are seasonally monoestrous, and their hormonal regulation patterns tend to be very similar to those of hibernating bats [33, 34]. In these animals, the breeding period appears to be confined to only one period of the year, with the testicular reactivation and spermatogenesis occurring prior to this period, and a process of testicular regression occurring later [33, 34]. However, variations to this pattern have been postulated. Dwyer [43] demonstrates two male reproductive peaks, with a reduced or regressed period between them, in *Myotis adversus* from

Australia, while Myers [44] postulates that *M. albescens* breeds two or three times annually, with spermatogenesis occurring only from autumn to spring in Paraguay.

Similarly, studies on *M. nigricans* show that its annual reproductive cycle is geographically variable, with specimens from Paraguay presenting an active pattern throughout the year; those from Panama exhibiting an active pattern for most of the year, but becoming reproductively quiescent for about three months – testicular regression (September to November); animals from Mexico resembling temperate zone bats in their reproductive patterns [34, 45]; and animals from the northwest of São Paulo State, Brazil showing two peaks of spermatogenic activity, followed by two periods of testicular regression, in the same annual reproductive cycle [9].

The observation of one period of testicular regression in the annual reproductive cycle of hibernating, and also in some nonhibernating, species of bats is common, but the observation of two periods in the same annual reproductive cycle seems to be specific to *M. nigricans* [9].

The annual reproductive cycle of *M. nigricans* from the northwest of São Paulo (Brazil) can be subdivided into four different periods (active, regressing, regressed and recrudescence), which repeat in the same order twice during the year [9, 42]. The two peaks of spermatogenic activity occur during the active period and processes of testicular regression follow both, from the regressing to the regressed periods [9]. Beguelini and collaborators [9, 12] demonstrated that this pattern is regulated by T and E2, *via* variations in the production of T, its conversion to E2, and the activation/ deactivation of the AR in Sertoli cells and the ERs in germ cells (spermatogonia, elongated spermatids and sperm).

In periods of recrudescence, an increase in the expressions of T and AR triggers the testicular reactivation and the restart of the spermatic production. The continuous increase in the expressions of both T and AR leads to the peak of spermatogenic activity in the active period. After the breeding period, the expressions of T and AR tend to decrease sharply, which causes the progressive testicular regression (regressing – regressed periods) [9, 12].

The deactivation of testicular production of T in regressing periods also affects the epididymal morphophysiology, causing a regression in its epithelium. However, the epididymis experiences a delay in regression, with it already having an active pattern in the regressing period and being regressed only in the regressed period. The maintenance of epididymal morphophysiology in the regressing period seems to be related to the presence of spermatozoa (sperm storage) and other testicular factors in the luminal compartment, with the luminal T secreted by the testes together with the spermatozoa being used to stimulate/regulate epididymal cells [18].

Moreover, the decrease in the testicular production of T in regressing periods directly affects the amount of circulating T available to stimulate the prostate. As a result, there is not enough circulating T in this period to stimulate prostatic expression and secretion. Thus, the process of testicular regression induces a concomitant process of prostatic regression [19].

In the recrudescence periods, the increase in the production of T increases the amount of circulating T, which could lead to the reactivation of the epididymis and prostate. However, the testicular recrudescence process is not followed by the reactivation of the epididymal epithelium, which presents low expressions of AR and ER in this period, showing that it is only possible that the epididymis recovers its full morphofunctional activity when spermatozoa and a great amount of T reach its lumen [18]. Similarly, despite the apparent physiological reactivation, with the high expressions of AR and ER, prostatic morphology alters little during this period, showing that the reactivation of the prostate is possibly delayed compared to that of the testes [19].

Interesting to note that these hormones (FSH, LH, T and E2) are not the only ones to act on the reproduction. Although there are few detailed studies, results indicate that other hormones such as leptin, melatonin, *etc.* may influence reproductive aspects of bats, such as seasonality, hibernation, and hibernation-related adaptations [46 - 51].

Studies indicated that leptin levels influences/regulates the metabolism and the feeding behavior, with high levels of circulating leptin tending to reduced food consumption; and impaired leptin signaling causing obesity [51]. Thus, hibernating bats must somehow to mediate or impair leptin expression during the pre-hibernation period, in order to accumulate sufficient energy for hibernation.

Similarly, it is well known that the main role of melatonin is to influence/control seasonal and diurnal biological rhythms [46 - 51]. Although this characteristic in itself demonstrated an important impact on the reproduction of bats, with melatonin being able to act regulating patterns of seasonality and hibernation, Singh and Krishna [49] demonstrated, in the bat *Scotophilus heathi*, that melatonin had inhibitory effects on testosterone production in the testes during winter dormancy, which may be the reason for the decline in testicular activity during winter.

Thus, we perceive that the responses of reproductive organs to annual seasonal variations depend not only on abiotic fluctuations, but mainly on an intricate relation of hormonal and signaling pathways that culminate in a better way of obtaining and using metabolic energy associated with the development of better forms of reproductive success.

PERSPECTIVES

Despite the huge number of species and unique reproductive adaptations that bats present, there are still many gaps in our knowledge of the hormonal regulation in male reproduction of Chiroptera. Thus, new studies that evaluate in depth the endocrine and paracrine hormonal regulation of male reproductive organs, in the face of the seasonal changes that they undergo during the annual cycle, are necessary and important in order to better understand the amazing reproductive adaptations that these animals have.

CONFLICT OF INTEREST

This research was supported by Grants 2009/16181–9 and 2009/03470–2 from the São Paulo State Research Foundation (FAPESP) and by Brazilian National Research and Development Council (CNPq) Processes: 300163/2008-8 and 301596/2011-5 research fellowships to SRT and Process: 302008/2010-1 to EMV. The authors disclose no conflicts of interest.

ACKNOWLEDGEMENTS

Mateus R. Beguelini contributes to all data analysis/interpretation and to the draft version and development of the chapter; Sebastião R. Taboga and Eliana Morielle-Versute critically revised and approved the chapter. The authors thank Luiz Roberto Falleiros Junior for technical help.

REFERENCES

[1] Holdcraft RW, Braun RE. Hormonal regulation of spermatogenesis. Int J Androl 2004; 27(6): 335-42. [http://dx.doi.org/10.1111/j.1365-2605.2004.00502.x] [PMID: 15595952]

[2] Kawamoto K. Endocrine control of the reproductive activity in hibernating bats. Zool Sci 2003; 20(9): 1057-69. [http://dx.doi.org/10.2108/zsj.20.1057] [PMID: 14578565]

[3] Krishna A, Bhatnagar KP. Hormones and Reproductive Cycles in Bats. In: Norris DO, Lopez KH, Eds. Hormones and Reproduction of Vertebrates. Mammals. Academic Press 2010; Vol. 5.

[4] Rannikki AS, Zhang FP, Huhtaniemi IT. Ontogeny of follicle-stimulating hormone receptor gene expression in the rat testis and ovary. Mol Cell Endocrinol 1995; 107(2): 199-208. [http://dx.doi.org/10.1016/0303-7207(94)03444-X] [PMID: 7768331]

[5] Sharpe RM. Regulation of Spermatogenesis. In: Knobill E, Neill JD, Eds. The Physiology of Reproduction. New York: Raven Press, Ltd 1994.

[6] O'Donnell L, Meachem SJ, Stantom PG, McLachlan RI. Endocrine Regulation of Spermatogenesis. In: Neill JD, Ed. Physiology of Reproduction. Academic Press 2006. [http://dx.doi.org/10.1016/B978-012515400-0/50026-9]

[7] Lei ZM, Mishra S, Zou W, *et al.* Targeted disruption of luteinizing hormone/human chorionic gonadotropin receptor gene. Mol Endocrinol 2001; 15(1): 184-200. [http://dx.doi.org/10.1210/mend.15.1.0586] [PMID: 11145749]

[8] Beguelini MR, Puga CC, Taboga SR, Morielle-Versute E. Annual reproductive cycle of males of the flat-faced fruit-eating bat, *Artibeus planirostris* (Chiroptera: Phyllostomidae). Gen Comp Endocrinol 2013; 185: 80-9. [http://dx.doi.org/10.1016/j.ygcen.2012.12.009] [PMID: 23356978]

[9] Beguelini MR, Góes RM, Taboga SR, Morielle-Versute E. Two periods of total testicular regression are peculiar events of the annual reproductive cycle of the black *Myotis* bat, *Myotis nigricans* (Chiroptera: Vespertilionidae). Reprod Fertil Dev 2014; 26(6): 834-46. [http://dx.doi.org/10.1071/RD13109] [PMID: 23830483]

[10] O'Donnell L, Robertson KM, Jones ME, Simpson ER. Estrogen and spermatogenesis. Endocr Rev 2001; 22(3): 289-318. [http://dx.doi.org/10.1210/er.22.3.289] [PMID: 11399746]

[11] Hess RA. Estrogen in the adult male reproductive tract: a review. Reprod Biol Endocrinol 2003; 1: 52. [http://dx.doi.org/10.1186/1477-7827-1-52] [PMID: 12904263]

[12] Beguelini MR, Falleiros LR Jr, Góes RM, Rahal P, Morielle-Versute E, Taboga SR. Differential expression of aromatase, estrogen receptor alpha and 17β-HSD associated with the processes of total testicular regression and recrudescence in the bat *Myotis nigricans* (Chiroptera: Vespertilionidae). Gen Comp Endocrinol 2014; 201: 53-64. [http://dx.doi.org/10.1016/j.ygcen.2014.03.044] [PMID: 24726986]

[13] Dorrington JH, Khan SA. Steroid production, metabolism and release by Sertoli cells. In: Russell LD, Griswold MD, Eds. The Sertoli Cell. Clearwater: Cache River Press 1993; pp. 538-49.

[14] Bagatell CJ, Dahl KD, Bremner WJ. The direct pituitary effect of testosterone to inhibit gonadotropin secretion in men is partially mediated by aromatization to estradiol. J Androl 1994; 15(1): 15-21. [PMID: 8188534]

[15] Hayes FJ, Seminara SB, Decruz S, Boepple PA, Crowley WF Jr. Aromatase inhibition in the human male reveals a hypothalamic site of estrogen feedback. J Clin Endocrinol Metab 2000; 85(9): 3027-35. [PMID: 10999781]

[16] Dupont S, Krust A, Gansmuller A, Dierich A, Chambon P, Mark M. Effect of single and compound knockouts of estrogen receptors alpha (ERalpha) and beta (ERbeta) on mouse reproductive phenotypes. Development 2000; 127(19): 4277-91. [PMID: 10976058]

[17] Eddy EM, Washburn TF, Bunch DO, *et al.* Targeted disruption of the estrogen receptor gene in male mice causes alteration of spermatogenesis and infertility. Endocrinology 1996; 137(11): 4796-805. [http://dx.doi.org/10.1210/endo.137.11.8895349] [PMID: 8895349]

[18] Beguelini MR, Góes RM, Rahal P, Morielle-Versute E, Taboga SR. Impact of the Processes of Total Testicular Regression and Recrudescence on the Epididymal Physiology of the Bat *Myotis nigricans* (Chiroptera: Vespertilionidae). PLoS One 2015; 10(6): e0128484. [http://dx.doi.org/10.1371/journal.pone.0128484] [PMID: 26057377]

[19] Beguelini MR, Góes RM, Rahal P, Morielle-Versute E, Taboga SR. Impact of the processes of testicular regression and recrudescence in the prostatic complex of the bat *Myotis nigricans* (Chiroptera: Vespertilionidae). J Morphol 2015; 276(7): 721-32. [http://dx.doi.org/10.1002/jmor.20373] [PMID: 25703716]

[20] Robaire B, Hamzeh M. Androgen action in the epididymis. J Androl 2011; 32(6): 592-9. [http://dx.doi.org/10.2164/jandrol.111.014266] [PMID: 21764895]

[21] Robaire B, Ewing LL, Zirkin BR, Irby DC. Steroid delta4-5alpha-reductase and 3alpha-hydroxysteroid dehydrogenase in the rat epididymis. Endocrinology 1977; 101(5): 1379-90.

[http://dx.doi.org/10.1210/endo-101-5-1379] [PMID: 913319]

[22] Felden F, Leheup B, Fremont S, *et al.* The plasma membrane of epididymal epithelial cells has a specific receptor which binds to androgen-binding protein and sex steroid-binding protein. J Steroid Biochem Mol Biol 1992; 42(3-4): 279-85. [http://dx.doi.org/10.1016/0960-0760(92)90130-B] [PMID: 1318734]

[23] Cunha GR, Donjacour AA, Cooke PS, *et al.* The endocrinology and developmental biology of the prostate. Endocr Rev 1987; 8(3): 338-62. [http://dx.doi.org/10.1210/edrv-8-3-338] [PMID: 3308446]

[24] Syntin P, Dacheux JL, Dacheux F. Postnatal development and regulation of proteins secreted in the boar epididymis. Biol Reprod 1999; 61(6): 1622-35. [http://dx.doi.org/10.1095/biolreprod61.6.1622] [PMID: 10570012]

[25] Hess RA, Bunick D, Lee KH, *et al.* A role for oestrogens in the male reproductive system. Nature 1997; 390(6659): 509-12. [http://dx.doi.org/10.1038/37352] [PMID: 9393999]

[26] Pearl CA, Berger T, Roser JF. Estrogen and androgen receptor expression in relation to steroid concentrations in the adult boar epididymis. Domest Anim Endocrinol 2007; 33(4): 451-9. [http://dx.doi.org/10.1016/j.domaniend.2006.09.003] [PMID: 17034985]

[27] Prins GS, Birch L, Greene GL. Androgen receptor localization in different cell types of the adult rat prostate. Endocrinology 1991; 129(6): 3187-99. [http://dx.doi.org/10.1210/endo-129-6-3187] [PMID: 1954898]

[28] Prins GS, Cooke PS, Birch L, *et al.* Androgen receptor expression and 5 alpha-reductase activity along the proximal-distal axis of the rat prostatic duct. Endocrinology 1992; 130(5): 3066-73. [http://dx.doi.org/10.1210/endo.130.5.1572313] [PMID: 1572313]

[29] Sánchez P, Torres JM, Castro B, Olmo A, del Moral RG, Ortega E. Expression of steroid 5α-reductase isozymes in prostate of adult rats after environmental stress. FEBS J 2013; 280(1): 93-101. [http://dx.doi.org/10 1111/febs.12052] [PMID: 23122426]

[30] Heldring N, Pike A, Andersson S, *et al.* Estrogen receptors: how do they signal and what are their targets. Physiol Rev 2007; 87(3): 905-31. [http://dx.doi.org/10.1152/physrev.00026.2006] [PMID: 17615392]

[31] Ellem SJ, Risbridger GP. The dual, opposing roles of estrogen in the prostate. Ann N Y Acad Sci 2009; 1155: 174-86. [http://dx.doi.org/10.1111/j.1749-6632.2009.04360.x] [PMID: 19250203]

[32] Reis NL, *et al.* Mamíferos do Brasil. Universidade Estadual de Londrina 2011.

[33] Crichton EG, Krutzsch PH. Reproductive biology of bats. London: Academic Press 2000.

[34] Krutzsch PH. Male reproductive patterns in nonhibernating bats. J Reprod Fertil 1979; 56(1): 333-44. [http://dx.doi.org/10.1530/jrf.0.0560333] [PMID: 381651]

[35] Krutzsch PH, Crichton EG. Observations on the reproductive anatomy of the male *Molossus fortis* (Chiroptera: Molossidae) with comments on the chronology of reproductive events. Mammalia 1990; 54: 287-96. [http://dx.doi.org/10.1515/mamm-1990-0212]

[36] Jolly SE, Blackshaw AW. Sex steroid levels and Leydig cell ultrastructure of the male common sheath-tail bat, Taphozous georgianus. Reprod Fertil Dev 1989; 1(1): 47-53. [http://dx.doi.org/10.1071/RD9890047] [PMID: 2710959]

[37] Singh UP, Krishna A. Seasonal changes in circulating testosterone and androstenedione concentration and their correlation with the anomalous reproductive pattern in the male Indian sheath-tailed bat, *Taphozous longimanus*. J Exp Zool 2000; 287(1): 54-61. [http://dx.doi.org/10.1002/1097-010X(20000615)287:1<54::AID-JEZ7>3.0.CO;2-Z] [PMID:

10861550]

[38] Gustafson AW. Male reproductive patterns in hibernating bats. J Reprod Fertil 1979; 56(1): 317-31. [http://dx.doi.org/10.1530/jrf.0.0560317] [PMID: 381650]

[39] Araújo RA, Amaro BD, Talamoni SA, Godinho HP. Seasonal reproduction of yellowish myotis, *Myotis levis* (Chiroptera: Vespertilionidae), from a Neotropical highland. J Morphol 2013; 274(11): 1230-8. [http://dx.doi.org/10.1002/jmor.20175] [PMID: 23907816]

[40] Bueno LM, Beguelini MR, Comelis MT, Taboga SR, Morielle-Versute E. Ultrastructure of spermatogenesis, spermatozoon and processes of testicular regression and recrudescence in *Eptesicus furinalis* (Chiroptera: Vespertilionidae). Anim Reprod Sci 2014; 148(3-4): 228-44. [http://dx.doi.org/10.1016/j.anireprosci.2014.05.018] [PMID: 24954586]

[41] Beguelini MR, Puga CC, Martins FF, Betoli AH, Taboga SR, Morielle-Versute E. Morphological variation of primary reproductive structures in males of five families of neotropical bats. Anat Rec (Hoboken) 2013; 296(1): 156-67. [http://dx.doi.org/10.1002/ar.22613] [PMID: 23117997]

[42] Beguelini MR, Taboga SR, Morielle-Versute E. Ultrastructural characteristics of the spermatogenesis during the four phases of the annual reproductive cycle of the black *myotis* bat, *Myotis nigricans* (Chiroptera: Vespertilionidae). Microsc Res Tech 2013; 76(10): 1035-49. [http://dx.doi.org/10.1002/jemt.22264] [PMID: 23857678]

[43] Dwyer PD. Latitude and breeding season in a polyestrus species of *Myotis*. J Mammal 1970; 51(2): 405-10. [http://dx.doi.org/10.2307/1378506]

[44] Myers P. Patterns of reproduction of four species of vespertilionid bats in Paraguay. Univ Calif Publ Zool 1977; 107: 1-41.

[45] Wilson DE, Findley JS. Reproductive cycle of a neotropical insectivorous bat, *Myotis nigricans*. Nature 1970; 225(5238): 1155. [http://dx.doi.org/10.1038/2251155a0] [PMID: 5418251]

[46] Anthony EL. Endocrinology of reproduction in bats: central control. In: Crichton EG, Krutzsch PH, Eds. Reproductive Biology of Bats. New York: Academic Press 2000; pp. 1-26. [http://dx.doi.org/10.1016/B978-012195670-7/50002-3]

[47] Banerjee A, Meenakumari KJ, Udin S, Krishna A. Melatonin regulates delayed embryonic development in the short-nosed fruit bat, *Cynopterus sphinx*. Reproduction 2009; 138(6): 935-44. [http://dx.doi.org/10.1530/REP-09-0114] [PMID: 19690096]

[48] Beasley LJ, Smale L, Smith ER. Melatonin influences the reproductive physiology of male pallid bats. Biol Reprod 1984; 30(2): 300-5. [http://dx.doi.org/10.1095/biolreprod30.2.300] [PMID: 6704467]

[49] Singh K, Krishna A. Inhibitory effects of melatonin on testosterone but not on androstenedione production during winter in the vespertilionid bat, *Scotophilus heathi*. J Pineal Res 1995; 19(3): 127-32. [http://dx.doi.org/10.1111/j.1600-079X.1995.tb00181.x] [PMID: 8750346]

[50] Srivastava RK, Krishna A. Melatonin modulates glucose homeostasis during winter dormancy in a vespertilionid bat, *Scotophilus heathi*. Comp Biochem Physiol A Mol Integr Physiol 2010; 155(3): 392-400. [http://dx.doi.org/10.1016/j.cbpa.2009.12.006] [PMID: 20026417]

[51] Willis CK, Wilcox A. Hormones and hibernation: possible links between hormone systems, winter energy balance and white-nose syndrome in bats. Horm Behav 2014; 66(1): 66-73. [http://dx.doi.org/10.1016/j.yhbeh.2014.04.009] [PMID: 24768718]

CHAPTER 3

Relationship Between Subcutaneous Fat and Reproductive Activity in Males of *Leptonycteris yerbabuenae* in *Los Laguitos* Cave, Chiapas, Mexico

Matías Martínez-Coronel[1,*], Yolanda Hortelano-Moncada[2], Violeta Corral[1] and Laura Cuevas[1]

[1] *Departamento de Biología, UAM-Iztapalapa. Av. San Rafael Atlixco 186, Col. Vicentina, Iztapalapa, 09340, D.F., México*

[2] *Departamento de Zoología, Instituto de Biología, Universidad Nacional Autónoma de México, 3er. Circuito Exterior S/N, Cd. Universitaria, Coyoacán, Ciudad de México, 04510, Apartado Postal 70-153, México*

Abstract: *Leptonycteris yerbabuenae* is a bat with a distribution range from southern USA to El Salvador. A distinctive feature of the species is the seasonal accumulation of subcutaneous fat. Since this specie does not hibernate, it has been speculated that this energy reserve is used for migration and reproduction or when food availability decreases. As the role of fat in populations of southern Mexico is unknown, we carried out this study to determine the fat accumulation pattern and the role it plays during an annual cycle in a resident colony of this taxon. Results show that males follow a seasonal pattern, which begins with fat accumulation in February and peaks in May, right when the mating period begins. After this period, the amount of fat steadily decreases and disappears during November-January. We also found that the body and testicular mass are positively related with the amount of accumulated fat and with the development of an interscapular dorsal patch. Data suggest that stored fat is used for reproductive purposes and as an energy reserve to stay longer in the breeding colony, and that the interscapular dorsal patch may be taken as the evidence of reproductive success.

Keywords: Body mass, Dorsal patch, Fat pattern, Mating behavior, Subcutaneous fat, Testicular mass.

Migration and reproduction are energetically costly phenomena for any organism; therefore, the accumulation of fat as an energy reserve is a common strategy in

* **Corresponding author Matías Martínez-Coronel:** Departamento de Biología, UAM-Iztapalapa. Av. San Rafael Atlixco 186, Col. Vicentina, Iztapalapa, 09340, D.F., México; Tel: +55-6668-5736; E-mail: marti17@hotmail.com

Edith Arenas-Ríos (Ed.)

migratory mammals and birds [1 - 3]. In the case of migration, when the distance to travel is hundreds of kilometers nonstop, the amount of fat must be accurate, since a greater or lesser amount than what is precisely needed can mean failure for the migrant, as excess weight or lack of energy will prevent the organism to reach the desired location [4]. *Leptonycteris yerbabuenae* is a phyllostomid bat (Phyllostomidae) with a distribution range from southern USA to El Salvador [5]. Previous studies have reported that this species accumulates fat in certain seasons [6 - 9] and since it does not hibernate, speculations have been made that this energy reserve might be used for migration, reproduction or when food availability is scarce [6, 7, 10].

It has been documented that some specimens of this taxon weigh more than 50% of the normal weight, which represents an excessive burden for a migrant [10]. It is also known that males accumulate more fat than females [6, 7]. If it is assumed that fat is used as fuel for migration, then why both sexes do not store energy in the same proportion? A partial answer to these differences was given by Fleming and collaborators [6, 10] who reported that females in Chamela, Jalisco, travel up to northern Sonora and Arizona, covering a distance of at least 1500 km, a journey that takes more than two months; as for the males, their fate is unknown. In a subsequent study in the same colony, Stoner *et al.* [11] found evidence that not all females migrate as Fleming and coworkers originally proposed [6, 10, 12]. A different migratory behavior from the Chamela population was found in colonies of this species in central and southern Mexico, which behave as resident or regional migrants, though individuals of both sexes accumulate fat seasonally [7 - 9, 13].

Different studies have shown that fat accumulation in *L. yerbabuenae* starts before the mating period and reaches its maximum value at its very beginning, thereafter, fat content gradually decreases in the population [6 - 8]. Also, it has been observed that during the reproductive period, males of *L. yerbabuenae* and *L. curasoae* developed a glandular complex or dorsal patch (DP) [14, 15], the formation of which involves a high energy investment, as males with DP had lower body mass compared to individuals without it [16]. If this was the case, the development of this glandular complex could explain the differences in the amount of fat stored between the sexes of L. *yerbabuenae*. But it is also considered that the fat accumulated seasonally by *L. yerbabuenae* might work as a source of energy used for courtship during the breeding season [7].

Since the role of accumulated fat in males of *L. yerbabuenae* is still not clear, this paper gives the results of observations on the role of fat accumulated by males of this species in a population from southern Mexico.

STUDY AREA

Los Laguitos Cave is located 4 km NW of Tuxtla Gutierrez, Chiapas, México, (16° 46' 42" N; 93° 08' 55" W), at 781 m.a.s.l. The climate of the region is warm-subhumid, with a wet season from May to October, a well-marked intraestival drought during the rainy season in July and August, and a dry season the rest of the year. The annual mean temperature is 25.3°C [17]. The dominant vegetation around the cave corresponds to a deciduous seasonal forest [18].

METHODOLOGY

To capture males of *L. yerbabuenae*, a mist net was placed at the cave entrance for two consecutive nights each month between April 2009 and March 2010, except between May and July as indicated below. The net was placed perpendicular to the outflow of bats, for an average 5 seconds period, given the high density of bats emerging along with *L. yerbabuenae* from the cave. After this, the net was placed parallel to the outflow, in order to recover captured specimens. This operation was conducted every 15 minutes during the first two hours after sunset. Age, forearm length, body mass, fat index, reproductive status and DP presence were recorded for each *L. yerbabuenae* male captured. Forearm length was measured with a Mitutoyo digital vernier to the nearest 0.01 mm; body mass was measured with a spring balance to the nearest 0.1 g. Every individual with completely ossified phalanges and brown fur was considered adult [19]. For fat index, we used the five categories defined by Ceballos *et al.* [6]. In the case of reproductive phenology, testicular position was considered, among which three categories were recognized: males with nonpalpable testes in the groin or scrotum, males with inguinal testes, and males with scrotal testes. Testis length and width inside the scrotum were measured with a digital Vernier, both in individuals with scrotal testes and in specimens with inguinal testes that could be lowered into the scrotum. For testicular mass measurement, the procedure used by Wilkinson and McCracken [20] was followed: First, testicular volume was recorded, according to a prolate spheroid formula: (4/3)Πr2 = 0.5236 x length x width2 = 0.021 log10 testicular mass + 1.024 log10 testicular volume; then, testicular mass was estimated with the formula: log10 testis mass = 0.021 + 1.024 log10 testis volume. The combined mass of both testes was used in the analysis.

To determine the integration of the breeding colony, the cave was visited daily from 15 May 2009. Once the breeding colony was detected, visits were made every fifth day, from 1 June to 14 July 2009. The size of the breeding colony was estimated based on infrared video recordings with a Sony TRV 280 camera. Three counts were carried out: one at 21:00 h, another at 00:00 h and the last one at 03:00 h the next day. The sex ratio was estimated based on a sample obtained with a spoon net at 00:00 h one day before the recording. For each individual captured, only sex was determined and then it was immediately released. Also, between 21:30-22:30 h video recordings were made at a fixed point to detect copulation frequency. The video was subsequently analysed on a computer with Sony Vegas Pro 9.0 software (Sony Creative Software Inc., Madison, WI, USA).

DATA ANALYSIS

Mean and standard errors were calculated for every quantitative variable. To determine the degree of association between variables, the Pearson correlation index was estimated. Finally, an analysis of variance was used to determine differences in body mass and testicular mass between the different categories used for males [21].

RESULTS

The following results correspond to the information obtained from 198 adult male of *L. yerbabuenae* collected at the entrance of the cave.

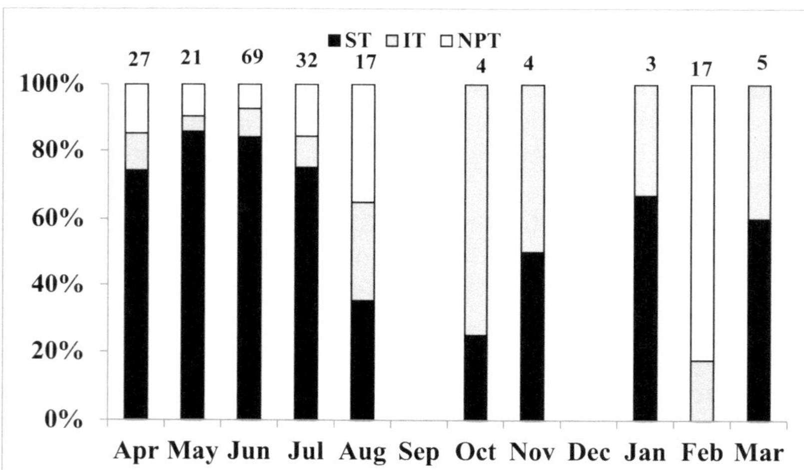

Fig. (1). *Leptonycteris yerbabuenae* male reproductive phenology at *Los Laguitos* Cave, Chiapas, México, between April 2009 and March 2010. ST, individuals with scrotal testes; IT, individuals with inguinal testes; NPT, individuals with nonpalpable testes. Samples sizes for each month are displayed above the bar.

Reproductive Phenology

Males with no palpable testes were found between February and August, while individuals with inguinal or scrotal testes were found throughout the year (Fig. **1**). Between April and July, more than 70% of males had scrotal testes; this increase coincided with the mating season.

Relationships between Body Mass, Testicular Mass and Fat Index (FI)

Average body mass was 25.01 ± 0.22 g and showed a seasonal variation. It increased from April to May, when it reached its maximum value which coincided with the beginning of the mating season (Fig. **2**). From June onwards, body mass declined steadily until February, when it reached its minimum value. FI had a similar behavior, which also showed a positive relationship with body mass ($r = 0.76$, $F_{1, 167\ df} = 278.20$, $P < 0.0001$) throughout the period studied.

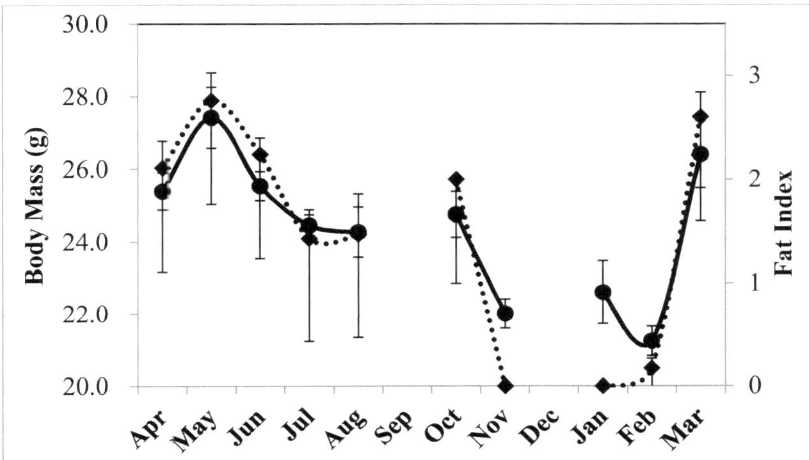

Fig. (2). Relationship between body mass (—◆—) and fat index (···◆···) ($r = 0.76$, $F_{1, 167\ df} = 278.20$, $P < 0.0001$) in *L. yerbabuenae* males in *Los Laguitos Cave*, Chiapas, Mexico, from April 2009 to March 2010.

Testicular mass also varied seasonally (Fig. **3**), increasing from April to May, when it reached its maximum value ($x = 138.04 \pm 19.23$ mg), coinciding with the beginning of the mating season. It began to decline from June onwards and this trend continued in the following months reaching its lowest value in February ($x = 7.56 \pm 7.01$ mg). The association between testicular mass and FI was moderate ($r = 0.46$, $R^2 = 0.21$, $F_{1, 171\ df} = 47.68$, $p = 0.0000$).

Dorsal Patch Development

The interscapular glandular complex or dorsal patch (DP) was developed exclusively by males between May and August. From this condition, two categories were recognized: one represented by individuals who had the back covered with oil, or sebaceous dorsal patch (SDP), and the other represented by individuals with an oily back and a hairless spot in the center of the patch, *i.e.* bare dorsal patch (BDP). We also made distinction of individuals that did not develop the glandular complex, *i.e.* without dorsal patch (WDP).

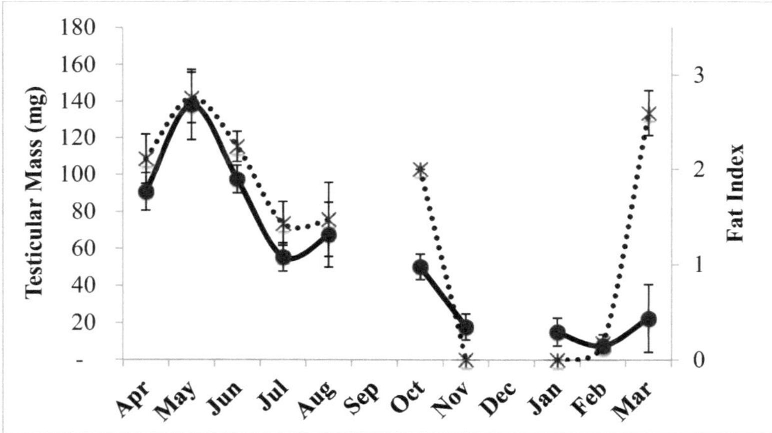

Fig. (3). Relationship between testicular mass (—●—) and fat index (···◆···) ($r = 0.46$, $R^2 = 0.21$, $F_{1,171df} = 47.68$, $p = 0.0000$) in *L. yerbabuenae* males in *Los Laguitos* Cave, Chiapas, Mexico, from April 2009 to March 2010.

The appearance of DP coincided with the beginning of the species mating season, which began in late May and ended in late June, and its presence was detected until August, two months after the end of this event. In May, 61.9% of males were placed in the category SDP; thereafter, their representativeness diminished down to 11.7% in August. On the other hand, BDP individuals were detected for the first time in May (4.7%) and in the following months their presence increased and raised up to 53.1% in July and 41.1% in August (Fig. **4**). Individuals in the category WDP were present in all months studied, accounting for 100% in April, as well as from October to March.

To determine whether the body mass of an individual was associated or not with the development of a DP, during the period from May to August, we compared the above three categories but the differences were not significant ($F_{1, 139\ df} = 2.47$, $P = 0.08$). Nonetheless, SPD individuals had higher body mass (25.90 ± 0.40 g) than BDP (25.54 ± 0.51 g) or WDP (24.41 ± 0.54 g) individuals. However, when analyzing samples per month, significant differences between these categories ($F_{1, 68\ df} = 4.60$, $p = 0.01$) were detected only in June, but not in May ($F_{1,20\ df} = 0.13$, $p = 0.72$), July (F1,31 df = 1.40, p = 0.26) or August (F1,16 df = 2.35, p = 0.13). In general, we found that body mass of individuals in the three categories tended to decrease from May to August (Fig. **5**).

When comparing testicular mass of individuals in the three DP categories, significant differences were found ($F_{1, 121\ df} = 5.57$, $p = 0.004$) in the period from May to August. SDP individuals had the highest testicular mass (108.55 ± 9.51 mg) compared with BDP (87.60 ± 8.83 mg) or WDP (59.50 ± 9.59 mg) categories, belonging the extreme categories SDP and WDP to different groups, according to an *a posteriori* Tukey-Cramer test. Also, a steady decline in testicular mass was observed from May to July in the analyzed categories (Fig. **6**).

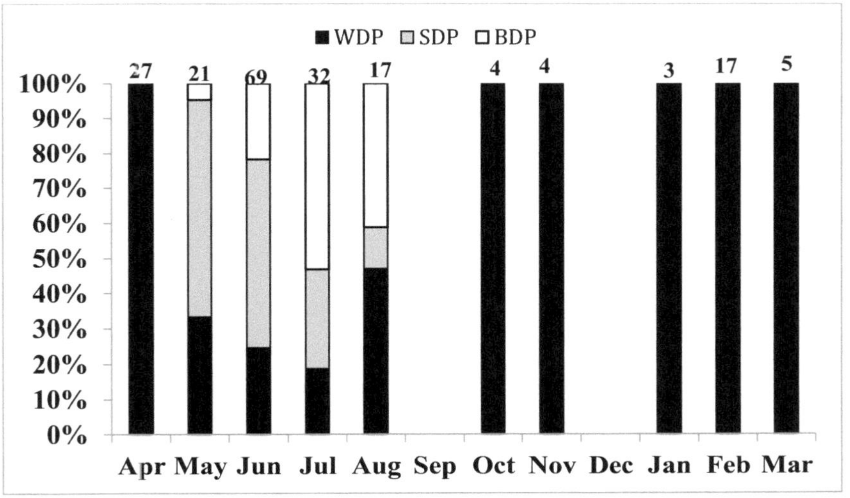

Fig. (4). Dorsal patch phenology in male *L. yerbabuenae* in *Los Laguitos* Cave, Chiapas, Mexico, present from April 2009 to March 2010. BDP, individuals with bare dorsal patch; SDP, individuals with sebaceous dorsal patch; WDP, individuals without dorsal patch. Samples sizes for each month are displayed above the bar.

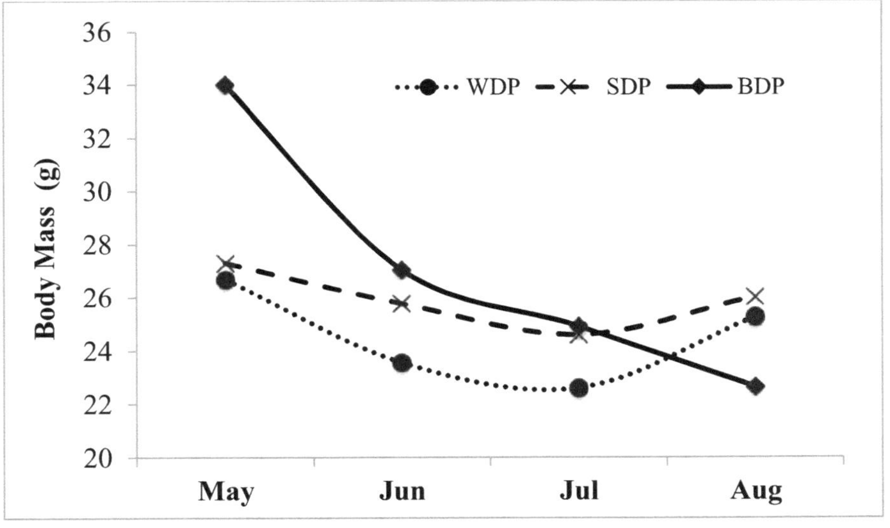

Fig. (5). Body mass changes in male *Leptonycteris yerbabuenae* present between May and August 2009, in *Los Laguitos* cave, Chiapas, Mexico. WDP, individuals without dorsal patch, SDP, individuals with sebaceous dorsal patch, BDP, individuals with bare dorsal patch.

Breeding Colony

The next information comes from the breeding colony detected in the interior of the cave, that was integrated by 204 adults, 121 males and 83 females.

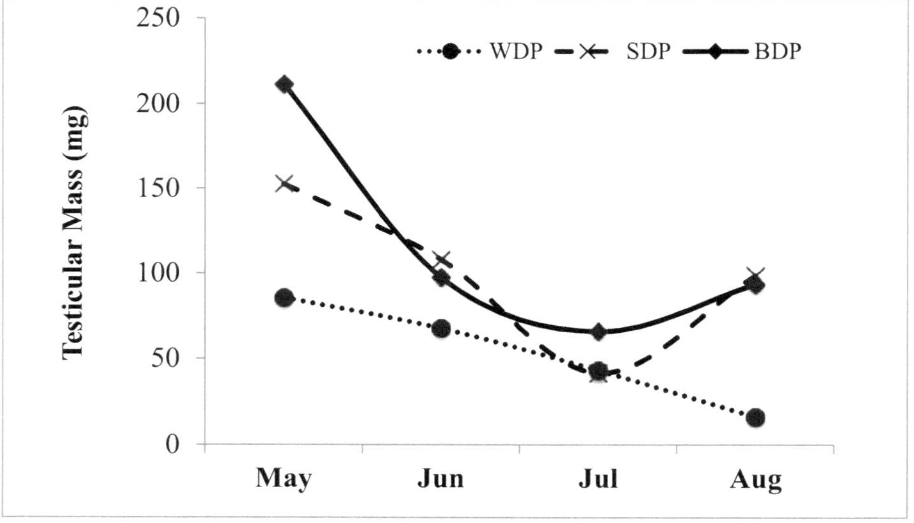

Fig. (6). Testicular mass pattern in male *Leptonycteris yerbabuenae* present between May and August 2009, in *Los Laguitos* cave, Chiapas, Mexico. WDP, individuals without dorsal patch, SDP, individuals with sebaceous dorsal patch, BDP, individuals with bare dorsal patch.

The breeding colony integrated on 30 May 2009 and disintegrated between 5 and 7 July 2009. During the mating period the colony showed changes in abundance during the night, and on average there were more individuals at 21:00 h (158.37 ± 87.15) and 3:00 h (141.28 ± 63.94), than at 00:00 h (116.62 ± 64.65). Males outnumbered females at a 1.4:1 ratio in the breeding colony.

Along eight hours of video 52 heterosexual intercourses were observed, with an average duration of 65.75 ± 5.09 s (11-146 s). Also, an intercourse that lasted 300 s was observed, which was excluded from the statistical data since it was considered unusual as it lasted more than double the maximum recorded of 146 s. We also detected four attempts of heterosexual intercourse and six attempts of intercourse between males, one of which was successful, with duration of 40 s. Likewise, from 53 heterosexual events registered, in 10 cases (18.86%) the coupling male was bitten by external males right in the dorsal patch, and only once a female did it when the intercourse was over.

DISCUSSION

The presence of individuals of *L. yerbabuenae* with scrotal testes all year around in *Los Laguitos* Cave should not be taken as an indication that they are reproductively active and having mature sperm. A previous study reported that this species has only one spermatogenic period throughout the year and that it lacks epididymal sperm storage capacity [15]. It is also known that there is a positive relationship between sperm maturation and testis size for spermatogenesis to occur [8, 15]. Therefore, the presence of scrotal testes in this species should not be considered as a signal of reproductive activity, unless it is combined with testicular mass or size.

The observed changes in body mass of *L. yerbabuenae* in *Los Laguitos* can be explained by the variation in the amount of subcutaneous fat. Fat accumulation has been documented in different species of bats; depending on the species, they use it for different purposes: as an energy reserve for hibernation, migration, reproduction, or to solve seasonal food scarcity [22 - 24]. The presence of subcutaneous fat in *L. yerbabuenae* was originally reported for a population in Chamela, Jalisco, Mexico [6]. These authors found that males of this species have fat all year round, in greater amount during the wet season (July-October) than in the dry season, with a peak in November-December. They also assumed that fat was used for migration, since in December this population leaves their day-time refuge. Subsequently, was reported that in a population from Tzinacanostoc, Puebla, Mexico, the absence of fat was evident in some months of the dry season (January-April), but during the wet season, a maximum was reached in July [7]. They also found that males migrated, but only a distance of 7 km from the previous site, and concluded that fat was not used for migration. Therefore, they wondered about the role of fat in this population, since given the geographical location of the cave, food resource availability was not an important variable to explain such an accumulation, so they suggested that this reserve should be used for reproductive purposes.

In the *L. yerbabuenae* population in *Los Laguitos,* we found that at the beginning of the dry season (November-January) males lack subcutaneous fat, but from February on it begins to accumulate and reaches a peak in May, coinciding with the beginning of the mating period. Copulations in *Los Laguitos* occur one month earlier than in Puebla, Mexico, and five months before than in Chamela, Jalisco, Mexico [6, 7]. First, we found that temporary differences in fat accumulation between these populations reflect a latitudinal effect and the already recognized segregation between populations above parallel 21°N and southern populations, which differ by having a different migratory and reproductive behavior [9, 13].

In the Chamela and Tzinacanostoc populations males migrate after the mating period [6, 7], however, in *Los Laguitos,* they remain as a resident population. Notwithstanding, and as in the other populations, body and testicular mass decreased in the months after mating. The decrease in body condition is neither satisfactorily explained by the lack of food resource, nectar and pollen. Since in the study area, the dominant vegetation corresponds to a deciduous seasonal forest, a habitat characterized by two peaks of food abundance, one of which occurs at the end of the dry season and the beginning of the rainy season [11, 25, 26] just when mating is carried out in this population.

Decrease in body condition of *L. yerbabuenae* males in *Los Laguitos* is not related to migration or lack of food, data suggest that fat accumulated by this species is used for breeding purposes and might relate to the energy expenditure implied by the night stay in the breeding colony. More males than females (1.4:1) were found in the breeding colony, suggesting a differential stay between sexes during this time. We assume that females only integrate to the colony for the time needed and once mounted, they do not copulate anymore, as they have a commitment to the future cub [27]. Males, instead, stay longer and try to mount as many females, and some nights they even do not go out to forage, as in other mammalian species [28], which ultimately leads to a decrease in body condition. Consequently, the energy cost might be higher for them than for females at this time, as happens in other bat species such as *Myotis daubentonii*, *Hypsugo savii* [29], *Myotis nattereri* [30] and *Pteropus poliocephalus* [31]. In the case of *P. poliocephalus*, it has been documented that during the mating season, males increase their energy expenditure as a result of intrasexual competition, courtship, mating and defense of the harem [31]. Unlike this species, *L. yerbabuenae* does not form harems, neither has a stereotypical courtship behavior towards the female, but instead performs a forced intercourse, and we assume that the increased energy expenditure of the male must be related to their overnight stay in the breeding colony.

The dorsal patch (DP) is a glandular complex developed by *L. curasoae* and *L. yerbabuenae* males [14, 16, 32]. According to Muñoz-Romo and coworkers [16, 32, 33] the DP does not develop in all individuals, but only in those with greater testicular mass and lesser body mass, that also carry a lower burden of ectoparasites. So, these bats were more attractive to females in contrast with those without DP. In *L. yerbabuenae* at *Los Laguitos* it was found that the DP was developed by individuals with better body condition and greater testicular mass compared to WDP individuals. Data also indicate that the percentage of SDP members was higher in May and decreased by August, while the representativeness of BDP members was reversed, suggesting that the loss of hair in the middle of the DP is second to the appearance of the glandular secretion.

Also, the video observations reveal that abrasions and possibly the lack of hair in category BDP individuals are the result of aggression suffered by the coupling male, during, or at the end of mating. Therefore, the presence of a BDP could be considered as an indication of reproductive success as was previously suggested [16].

With regard to the mechanism on DP development, Muñoz-Romo and Kunz [16, 33] described a diurnal stereotypic behavior in males of *L. curasoae*, which was not detected during nocturnal observations in males of *L. yerbabuenae* in *Los Laguitos*. Failure to observe this behavior in this species does not mean it does not exist, but may be result of observations occurring at different times, as well as of different activities carried out by individuals during daytime *vs*. nighttime.

CONCLUSION

The results of this study reveal that males of *L. yerbabuenae* in *Los Laguitos* Cave show a temporal pattern in body and testicular mass values associated with the amount of subcutaneous fat. Fat began to accumulate in February and reached a peak in May; then, decreased in the following months and disappeared between November and January. Also, fat was used for reproductive purposes, not for migration or to cope with low food availability, and it is important as an energy reserve for an overnight stay of individuals in the breeding colony formed from late May to early July. Finally, we found that males with higher body and testicular mass were the ones that developed a dorsal patch, which may be an indication of reproductive success, as suggested by other authors.

CONFLICT OF INTEREST

The authors declare no conflict of interest, financial or otherwise.

ACKNOWLEDGEMENTS

To Marcelina Blas and Marisol Pérez Gutiérrez and families, who have always supported our stay in Tuxtla Gutiérrez. To Anubis Molina Gutiérrez, who supported us during the fieldwork. To Dr. Edith Arenas Rios for her invitation to participate in this volume.

REFERENCES

[1] Guglielmo CG. Move that fatty acid: fuel selection and transport in migratory birds and bats. Integr Comp Biol 2010; 50(3): 336-45. [http://dx.doi.org/10 1093/icb/icq097] [PMID: 21558208]

[2] McGuire LP, Guglielmo CG. What can birds tell us about the migration physiology of bats? J Mammal 2009; 90: 1290-7. [http://dx.doi.org/10.1644/09-MAMM-S-084R.1]

[3] Schneider JE. Energy balance and reproduction. Physiol Behav 2004; 81(2): 289-317. [http://dx.doi.org/10.1016/j.physbeh.2004.02.007] [PMID: 15159173]

[4] McWilliams SR, Guglielmo CG, Pierce B, Klaassen M. Flying, fasting, and feeding in birds during migration: a nutritional and physiological ecology perspective. J Avian Biol 2004; 35: 377-93. [http://dx.doi.org/10.1111/j.0908-8857.2004.03378.x]

[5] Arita HT, Humphrey SR. Revisión taxonómica de los murciélagos magueyeros del género *Leptonycteris* (Chiroptera: Phyllostomidae). Acta Zool Mex (ns) 1988; 29: 1-60.

[6] Ceballos G, Fleming TH, Chávez C, Nassar J. Population dynamics of *Leptonycteris curasoae* (Chiroptera: Phyllostomidae) in Jalisco, México. J Mammal 1997; 78: 1220-30. [http://dx.doi.org/10.2307/1383065]

[7] Galindo-G C, Sánchez-Q A, Quijano RH, Herrera-M LG. Population dynamics of a resident colony of *Leptonycteris curasoae* (Chiroptera: Phyllostomidae) in central Mexico. Biotropica 2004; 36: 382-91.

[8] Martínez-Coronel M, Munguía-Pérez AA, Arenas-Ríos E. Relationship between the amount of subcutaneous fat, testicular morphometry, epididymis and some sperm parameters in *Leptonycteris yerbabuenae* bat before, during and after mating. AVS 2015; 3: 22-7. [http://dx.doi.org/10.11648/j.avs.s.2015030101.14]

[9] Rojas-Martínez A, Valiente-Banuet A, Arizmendi MC, Alcántara-Eguren A, Arita HT. Seasonal distribution of the long-nosed bat (*Leptonycteris curasoae*) in North America does a generalized migration pattern really exist? J Biogeogr 1999; 26: 1065-77. [http://dx.doi.org/10.1046/j.1365-2699.1999.00354.x]

[10] Fleming TH, Nassar J. Population biology of the lesser long-nosed bat *Leptonycteris curasoae* in Mexico and Northern South America. In: Fleming TH, Valiente-Banuet A, Eds. Columnar Cacti and Their Mutualists: Evolution, Ecology, and Conservation. Tucson: The University of Arizona Press 2002; pp. 283-305.

[11] Stoner KE, O.-Salazar KA, R.-Fernández RC, Quesada M. Population dynamics, reproduction, and diet of the lesser long-nosed bat (*Leptonycteris curaso*ae) in Jalisco, Mexico: implications for conservation. Biodivers Conserv 2003; 12: 357-73. [http://dx.doi.org/10.1023/A:1021963819751]

[12] Wilkinson GS, Fleming TH. Migration and evolution of lesser long-nosed bats *Leptonycteris curasoae*, inferred from mitochondrial DNA. Mol Ecol 1996; 5(3): 329-39. [http://dx.doi.org/10.1111/j.1365-294X.1996.tb00324.x] [PMID: 8688955]

[13] Álvarez T, Sánchez-Casas N, Villalpando JA. Registro de los movimientos de *Leptonycteris yerbabuenae* en el centro de México. Anales Esc nac Cienc biol México 1999; 45: 9-15.

[14] Nassar JM, Salazar MV, Quintero A, *et al.* Seasonal sebaceous patch in the nectar-feeding bats *Leptonycteris curasoae and L. yerbabuenae* (Phyllostomidae: Glossophaginae): phenological, histological, and preliminary chemical characterization. Zoology (Jena) 2008; 111(5): 363-76. [http://dx.doi.org/10.1016/j.zool.2007.10.006] [PMID: 18602804]

[15] Rincón-Vargas F, Stoner KE, Vigueras-Villaseñor RM, Nassar JM, Chaves OM, Hundson R. Internal and external indicators of male reproduction in the lesser long-nosed bat *Leptonycteris yerbabuenae*. J Mammal 2013; 94: 488-96. [http://dx.doi.org/10.1644/11-MAMM-A-357.1]

[16] Muñoz-Romo M, Kunz TH. Dorsal patch and chemical signaling in males of the long-nosed bat, *Leptonycteris curasoae* (Chiroptera: Phyllostomidae). J Mammal 2009; 90: 1139-47. [http://dx.doi.org/10.1644/08-MAMM-A-324.1]

[17] Comisión Nacional del Agua (CONAGUA). Datos de la estación climatológica 7202 Normales 1951-2010 2016. Available at: www.smn1.conagua.gob.mx/climatologia/normales/estacion/Estaciones-Climatologicas.kmz

[18] Miranda F. La vegetación de Chiapas. 3rd ed., Tuxtla Gutiérrez: Consejo Estatal para la Cultura y las Artes de Chiapas 1998.

[19] Martínez-Coronel M, Cervantes FA, Hortelano-Moncada Y. Crecimiento postnatal y desarrollo del vuelo en el murciélago *Leptonycteris yerbabuenae* en Chiapas, México. Therya 2014; 5: 303-22. [http://dx.doi.org/10.12933/therya-14-173]

[20] Wilkinson GS. McCracken GF. Sperm competition. In: Kunz TH, Fenton MB, Eds. Bat Ecology. Chicago: The University of Chicago Press 2003; pp. 128-55.

[21] Zar JH. Biostatistical Analysis. 3rd ed., New Jersey: Prentice Hall 1996.

[22] Kunz TH, Wrazen JA, Burnett CD. Changes in body mass and fat reserves in prehibernating little brown bats (*Myotis lucifugus*). Ecoscience 1998; 5: 8-17. [http://dx.doi.org/10.1080/11956860.1998.11682443]

[23] McNab BK. Seasonal fat reserves of bats in two tropical environments. Ecology 1976; 57: 332-8. [http://dx.doi.org/10.2307/1934821]

[24] Weber NS, Findley JS. Warm-season changes in fat content of *Eptesicus fuscus*. J Mammal 1970; 5: 160-2. [http://dx.doi.org/10 2307/1378548]

[25] Bullock SH. La fenología de plantas en Chamela. In: Noguera FA, Vega Rivera JH, García Aldrete AN, Quesada Avendaño A, Eds. Historia Natural de Chamela. Universidad Nacional Autónoma de México 2002; pp. 491-8.

[26] Bullock SH, Solís-Magallanes JA. Phenology of canopy trees of a tropical deciduous forest in Mexico. Biotropica 1990; 22: 22-35. [http://dx.doi.org/10.2307/2388716]

[27] Clutton-Brock TH, Albon SD, Guinness FE. Fitness costs of gestation and lactation in wild mammals. Nature 1989; 337(6204): 260-2. [http://dx.doi.org/10.1038/337260a0] [PMID: 2911365]

[28] Miquelle DG. Why don't bull moose eat during the rut? Behav Ecol Sociobiol 1990; 27: 145-51. [http://dx.doi.org/10.1007/BF00168458]

[29] Rughetti M, Toffoli R. Sex-specific seasonal change in body mass in two species of vespertilionid bats. Acta Chiropt 2014; 16: 149-55. [http://dx.doi.org/10.3161/150811014X683363]

[30] Kohyt J, Rozik A, Kozakiewicz K, Pereswiet-Soltan A, Gubala WJ. Activity pattern and fat accumulation strategy of the Natterer's bat (Vespertilionidae, Chiroptera) swarming population indicate the exact time of male mating effort. Mammal Res 2016; 61: 149-89. [http://dx.doi.org/10.1007/s13364-016-0285-x]

[31] Welbergen JA. Fit females and fat polygynous males: seasonal body mass changes in the grey-headed flying fox. Oecologia 2011; 165(3): 629-37. [http://dx.doi.org/10.1007/s00442-010-1856-1] [PMID: 21153744]

[32] Muñoz-Romo M, Burgos JF, Kunz TH. The dorsal patch of males of the Curaçaoan long-nosed bat, *Leptonycteris curasoae* (Phyllostomidae: Glossophaginae) as a visual signal. Acta Chiropt 2011; 13: 207-15. [http://dx.doi.org/10.3161/150811011X578769]

[33] Muñoz-Romo M, Burgos JF, Kunz TH. Smearing behavior of male *Leptonycteris curasoae* (Chiroptera) and female responses to the odour of dorsal patches. Behaviour 2011; 148: 461-83. [http://dx.doi.org/10.1163/000579511X564287]

CHAPTER 4

Evidence of Apoptosis in the Testicular Blood Vessels of *Corynorhinus mexicanus* Bats Characterized by Seasonal Reproduction

Miguel Angel León-Galván[1,*], **Ahiezer Rodríguez-Tobón**[1] and **Rosa María Vigueras-Villaseñor**[2]

[1]*Departamento de Biología, Universidad Autónoma Metropolitana - Iztapalapa, Ciudad de México, México*

[2] *Laboratorio de Biología de la Reproducción, Instituto Nacional de Pediatría, Ciudad de México, México*

Abstract: In the species of bats that present a asynchronous reproductive pattern (Corynorhinus mexicanus), spermatogenesis is in recess during autumn and winter, when androgen levels are low, but the secondary sexual functions are maintained and mating takes place. The proliferation of spermatogonia and the spontaneous degeneration of spermatocytes often coincide during normal testicular functioning in mammals, and this is where apoptosis plays a central role in controlling the number of testicular germ cells. The spermatogonia are the cells that frequently suffer apoptosis in the testicles, but this process can also affect spermatocytes and spermatids. This emphasizes the need to improve our understanding of the participation of apoptosis and, therefore, its modulation during testicular development and training of new blood vessels. The strict regulation of angiogenesis and its adequate functioning are very important for organisms, since both the excessive formation and the insufficient development of blood vessels can cause severe diseases. Studies of the formation and involution of testicular blood vessels in seasonal reproductive bat species may thus shed new light on understanding the control and development of angiogenesis in mammals. Using the *C. mexicanus* bat as a study model of the seasonal activity of the testicle, we determine for the first time the involvement of the apoptosis in developing testicular vasculature in a seasonal chiropteran with drastic morphological changes of the testes, although more comparative studies are needed, mainly one between the different stages of testicular cycle which can provide information of the existence of angiogenesis associated with the testicular cycle in these species.

Keywords: Apoptosis, Bats, Blood vessels, Seasonal reproduction, Testicles.

[*] **Corresponding author Miguel A. León-Galván:** Department of Biology, Autonomous Metropolitan University-Iztapalapa, San Rafael Atlixco No. 186, Col. Vicentina, Iztapalapa, 09340, Ciudad de México, México; Tel: +55 5804 3049; E-mail: leon@xanum.uam.mx

Edith Arenas-Ríos (Ed.)

INTRODUCTION

The Order Chiroptera (bats) is the second-most diverse group in the Class Mammals, as it includes approximately 25% of all species described (around 1130). Moreover, bat populations are among the most varied as there are records of species with a size of just a few hundred individuals, while others may number in the hundreds of thousands, some of which are the most numerous in this Class. It is a cosmopolitan group as there are representatives in virtually every terrestrial ecosystem in the world except polar zones and some islands. Their diversity is reflected in an interesting variety of guilds distinguished by type of alimentation, beginning with a division into two broad categories: phytophagous (herbivores), that includes mostly frugivorous bats, with lower proportions of nectarivorous and pollinivorous species. The second category is made up of the carnivorous species, including mainly insectivores, but with some fish-eaters (piscivorous), others that feed on small vertebrates (such as frogs or birds), and the more specialized hematophagous bats whose only source of alimentation is the blood of mammals or birds, according to the species (only 3 found exclusively in tropical America).

Without doubt, another factor that greatly influences the success of Chiroptera, and one directly-associated with their richness is the broad variety of reproductive patterns they present. While bats follow the general reproductive pattern of eutherian mammals (*i.e.*, internal fertilization, viviparity, placentation), they have characteristically long gestation and lactation periods, which together last 5 to 6 months [1] compared to other mammals of proportionally similar body size, such as many rodents. Most bat species are monotocous.

REPRODUCTIVE PATTERNS IN BATS

A species' reproductive pattern is classified primarily on the basis of the number of estrous cycles and the seasonality with which females experience them. Using these criteria bats are grouped into two categories: polyestrous species that are divided, in turn, into two modalities called continuous and seasonal; and monoestrous species, most of which are seasonal. Regarding the males, activity patterns strictly follow the rhythm of the reproductive periodicity that the females of the species present. As a result, the reproductive patterns of male bats are classified as either continuous or seasonal, depending on the period of spermatogenic activity during the annual cycle.

Most polyestrous bats inhabit tropical regions. Characteristic examples are the hematophagous species *Desmodus rotundus*, which shows the continuous pattern [2, 3], and the frugivorous type *Artibeus lituratus*, which is bimodal-seasonal since it presents a main activity period in the spring and another, less important, in autumn [4]. However, the most generalized pattern among bat species for which there are sufficient knowledge of their reproductive biology is the monoestrous seasonal type, which occurs principally among insectivorous species living in temperate and cold regions of the world.

Considering the timing with which the three main male sexual functions –*i.e.*, spermatogenesis, sperm maturation and epididymal sperm storage and mating– occur, three reproductive modalities related to the synchrony among these sexual functions have been recognized for seasonal species in temperate zones [3, 5]. In the first, all functions occur synchronically during the only annual reproductive cycle that occurs. The second shows a degree of asynchrony, as spermatogenesis ends shortly after the onset of mating. The third case presents total asynchrony because spermatogenesis occurs in spring-summer, but mating does not begin until autumn (and in some species can extend into winter). This pattern requires a prolonged period of sperm storage in the cauda of the epididymis, with extensive secretory activity by the accessory sex glands at the time when mating take place. And this occurs with no new sperm production because spermatogenesis culminated some months before the beginning of the mating season.

Several studies have identified that the temporally-asynchronic seasonal reproductive pattern characterizes male bats belonging to a numerous group of species that permanently inhabit temperate and cold zones where they must resist adverse environmental conditions and low food availability during the winter months. This is achieved through their ability to sustain a condition of physiological lethargy that reduces their metabolism to allow energy savings as high as 90% compared to normal conditions of activity [6], but the consequence is that energy-consuming reproductive processes like spermatogenesis are suspended [7, 8].

This important group of species of temperate and cold zones around the world that present the asynchronous male reproductive pattern due to the effects of the winter season includes some members of the Rhinolophid and Vespertilionid families, and as part of this last family, all known species of the Plecotini tribe (Dobson, 1875), an holarctic group of long-eared bat species which has representatives in Eurasia (*Barbastella*, *Otonycteris* and *Plecotus* genera) where the group originated probably, and North America (*Corynorhinus*, *Euderma* and *Idyonicteris* genera) to where is distributed in the course of its evolution [9]. The genus *Corynorhinus* particularly, is comprised by three species, *C. rafinesquii* distributed only in United States of America; *C. townsendii* whose range include USA and Mexico; and *C. mexicanus*, an endemic species to Mexico.

Reproduction is controlled by several different external factors, including food availability, the timing of the rainy season, temperature and, especially, photoperiod. It is only when all these factors are in optimal conditions that successful reproduction can occur in these bat species [5, 10, 11].

EFFECT OF PHOTOPERIOD AND TEMPERATURE ON TESTICULAR FUNCTION

In most mammal species the reproductive season is controlled mainly by the hours of light. The altitude of the habitat also influences this, but less significantly. The photoperiod is a temporal information signal related to the initiation or termination of related morphological, physiological and behavioral changes that maximize species survival, including reproductive activity [12]. This affects the reproductive physiology of temperate zone bats in terms of the males' testicular weight. In the species *Antrouzous pallidus*, which has an asynchronous seasonal pattern, studies have determined that when live offspring are bred under laboratory conditions there is a direct relation between the weight of the testicles and more hours of light (*i.e.*, longer days), but that when fewer hours of light occur spermatogenic activity ceases.

This mechanism allows organisms to closely follow the annual rotation of seasons and anticipate the relatively predictable yearly variations in key environmental parameters. Information gathered through the retina is transduced into a physiological signal *via* the pineal hormone in a process that begins when a luminous stimulus that activates photoreceptors in the retina is transmitted to the suprachiasmatic nucleus *via* the optic nerve and then to the paraventricular nucleus. This photic information is then retransmitted by the intermediate-lateral cells of the upper spinal cord to the upper cervical ganglion, where the postganglionic innervation of the pineal gland regulates melatonin synthesis.

Melatonin secretion is related to circadian rhythms with high nighttime levels and basal secretion in daytime. Darkness promotes noradrenalin secretion by the upper cervical ganglion and its activity in the pineal gland increases the production of cyclic adenosine monophosphate (cAMP) and the incorporation of tryptophan, while also inducing the catalytic activation of N-acetyltransferase and hydroxide-O-methyltransferase, which result in the synthesis and release of melatonin. Variations in the length of the day can increase or decrease the sensitivity of the hypothalamus to the negative feedback of estradiol, which modifies the production of pulses of the gonadotropin-releasing hormone (GnRH), thus altering the secretion frequency of the luteinizing hormone (LH). Daylight duration is the primary signal for eliminating reproductive activity [13].

Therefore, temperature and the amount of daylight are two key environmental factors that can modify reproductive behaviors, though not all species are equally sensitive to these factors. Photoperiod directly influences ambient temperature; that is, greater illumination on the Earth's surface raises temperatures [14].

There is little information on the effects of high temperatures on the reproduction of wild mammals in natural habitats [15], but we do know that the testicles require a lower temperature than the normal body temperature in order to produce sperm, so temperature is considered a determining factor in male reproduction because of the energy costs involved [16]. Exposure to high temperatures has important implications for sterility and anti-conception in males, for a rise in the temperature of the male gonads increases the amount of reactive oxygen species (ROS), which can damage proteins which are essential to the signaling pathways that foster the later motility of gametes. Also, damage to membrane lipids causes oxidation of the fatty acids that form it (lipid peroxidation) and can affect the nucleic acids that make up the cells' DNA chains [17, 18].

Of course, body temperature is affected by ambient temperature. One way in which the testicles maintain their temperature below body temperature depends largely on the scrotum, which in some mammals constitutes an evolutionary adaptation. One function of the scrotum is to maintain the testicle at a certain distance from the body so that the male gonads remain at a temperature lower than the internal body temperature (37ºC). But mammals utilize two complex mechanisms to maintain temperatures around these values: a) regulating involuntary muscular movements; and, b) activating and deactivating various metabolic processes [19].

SPERMATOGENESIS IN BATS

Male mammal gonads are paired organs that have two main functions: sperm production and steroid synthesis. Sperm production is called spermatogenesis and occurs inside the seminiferous tubules. The interstitial space, meanwhile, contains Leydig cells (androgen producers), macrophages, blood, lymphatic channels and nerves. The intertubular cells are also accompanied by a network of blood vessels that allow the transit of many of the factors that participate in spermatogenesis, a process that can be divided into three main phases: proliferation, meiosis, and differentiation.

In the first phase, spermatogonia dispersed around the lumen of the seminiferous tubules proliferate by mitosis. Two types of spermatogonia are generally recognized: types A and B, though in some mammal species as many as 4 have been identified. Upon dividing, the type A spermatogonia produce the type B, which will then differentiate into primary spermatocytes through meiotic division (2n). The first complete meiotic division results in two secondary spermatocytes. Each secondary spermatocyte that enters meiosis II will produce two spermatids (n), which then differentiate to form spermatozoids. This differentiation constitutes the final phase of spermatogenesis, also known as spermiogenesis.

During spermiogenesis, the spermatids undergo a series of progressive structural and morphological changes that include: 1) formation of the flagellum of the spermatozoid from the centriolar region, together with the formation of a nucleus of microtubules; 2) condensation of the nuclear chromatin as histones are replaced by protamines; 3) movement of the nucleus towards the cell periphery; 4) development of the acrosomal cap to form the acrosome from the Golgi apparatus; and, 5) formation of the residual body through spillage of a large part of the cytoplasm that will be phagocyted by Sertoli cells.

Sertoli cells are somatic cells also found in the seminiferous epithelium, together with peritubular myoid cells. These cells form the wall of the seminiferous tubule and have characteristically been considered support cells that provide nutrients and growth factors to the germinal cells, but they perform the additional, important function of maintaining a characteristic intratubular environment by forming the hematotesticular barrier [20].

The Sertoli cells extend from the basal membrane to the lumen of the seminiferous tube. They are bonded by the basolateral membrane through tight junctions [20] formed by various transmembrane proteins including: claudin-11, occludin and the junctional adhesion molecule (JAM) [21]. The transmembranal proteins anchored to the cytoskeleton of Seroli cells through their interactions with cytosolic proteins [22]. In turn, the membrane of the Sertoli cells contains receptors and transporters in both the basolateral and apical portions that, together with the tight junctions, form the hematotesticular barrier. This barrier allows the generation of a microenvironment that is optimal for spermatogenesis, while delimiting the transit of toxic molecules from the interstitial space that could potentially affect spermatozoid production [23].

In primates and rats, a high level of testosterone and other factors produced by the Leydig cells are necessary for normal spermatogonia recruitment, while testosterone is recognized as playing a predominant role in spermiogenesis by stimulating maturation of the round spermatids until they become elongated [24]. The action of FSH is key to stimulating the progression of spermatogonia from type A to type B during the proliferative stage of spermatogenesis, and the secreting capacity of the Sertoli cells where it also affects the cytoskeletal arrangement required to maintain the hematotesticular barrier and cell adhesion [24]. The release of sperm from the seminiferous epithelium (called spermiation) is controlled synergistically by FSH and testosterone [24].

Many spermatogenesis-regulating factors travel through the bloodstream, and this led Yoshida *et al.* [25] to propose in 2007 that the niche in the germinal line in mammals is established as a result of the formation of vasculature patterns. Added to this is the microenvironment generated by the hematotesticular barrier that promotes the selective passage of elements that travel through the bloodstream.

MORPHOPHYSIOLOGICAL CHANGES IN MAMMAL TESTICLES ACCORDING TO SEASONAL REPRODUCTION

In seasonal mammals, including studied bat species, plasma testosterone concentrations are higher during the period of spermatogenesis [26], but bats show marked particularities; for example, a significant modification in the *Scotophilus heathi* (Vespertilionidae) from sub-tropical India. In this species, a high concentration of circulating androstenedione and the peak of plasma testosterone production are associated especially with the initial round of spermatogenesis [27]. But a totally distinct situation is reported for two species of Australian bats: the *Mormopterus planiceps* (Molossidae) and the *Taphozous georgianus* (Emballonuridae). Though not hibernating species, they have the capacity to greatly reduce their metabolism in winter. They have only one long cycle of spermatogenesis, and reports indicate that the onset of the cycle is marked by a coincident increase in plasma androstenedione and testosterone levels in *M. planiceps*, and in androstenedione, dihydrotestosterone and testosterone levels in *T. georgianus*, and that a small peak of testosterone coincides with the maximum size of the testicles in both species. However, testicular regression occurs despite the high concentration of these androgens [28]. The reproductive model that has drawn greatest attention is seen in some temperate and cold zones bats where the males' sexual functions present complete temporal asynchrony. One example is the *Corynorhinus mexicanus* that lives at high altitudes in Mexico [29]. In the species that present the asynchronous reproductive pattern, spermatogenesis is in recess during autumn and winter, when androgen levels are low, but the secondary sexual functions are maintained and mating takes place [3, 28].

In most species of insectivorous bats that inhabit temperate and cold zones, spermatogenesis corresponds to the peak availability of insects during half of the summer, and may be affected by corporal condition [30]. Seasonal variation in spermatogenesis involves a stage of recrudescence, another of profound involution of the seminiferous tubules and, in some cases, testicular migration towards the inguinal canal [3]. While androgenesis (at low levels) and the secondary reproductive functions are maintained during the winter [2, 28], spermatogenesis cannot proceed during the period in which the metabolic index characteristic of repose is persistently depressed and lethargy is prolonged. This causes a dissociation between spermatogenesis and hibernation [3, 28]. Even in tropical areas, spermatogenesis is interrupted or suppressed during the coldest, driest months [5]; and is delayed in summer when certain species of bats have recourse to daily torpor because of the marked reduction of ambient temperatures [8, 31, 32].

Temperate zone bats have longer periods of spermatogenesis than other mammals of similar size, as three to four months normally lapse between the first signs of growth in the seminiferous tubules, the release of sperm from the seminiferous epithelium, and the subsequent filling of the epididymis with mature sperm [3]. Given that during the period of reproductive activity some species of bats experience as much as a forty-fold increase in testicular mass [29], it has been proposed that the long duration of spermatogenesis may be due to the high energy costs required for growth of the testicles and the accessory sex glands [5]. Unfortunately, existing studies are insufficient to permit identification of the components of the testicle that determine this enormous increase in mass.

The principle changes observed in interstitial behavior between the active and inactive conditions of the testicles that have been reported for different species of seasonal mammals include modifications of the volume of complete tissue and of the blood vessels, Leydig cells and macrophages [33, 34], as well as dedifferentiation-redifferentiation [35], and changes in the number of Leydig cells [36, 37]. Regarding the tubular compartment, changes have been recorded in the volume, diameter and length of the seminiferous tubules [38], and in the size of the tubular lumen [34].

Sertoli cells have received special attention because of their intimate relation with the spermatogenesis process. The main seasonal variations in these somatic cells are morphological and functional in nature [39 - 41], for it is believed that the number of these cells is stable and unmodifiable by hormones after puberty [42, 43]. These cells are notable for their resistance to stimuli that induce death in other cell types [44, 45], and are considered terminally-differentiated in their adult form [30]. However, some studies in species with seasonal reproduction (horses [37, 46], hamsters [34, 47], vizcachas [41], roe deer [38] and rams [48, 49]) report the existence of seasonal variations in the number of Sertoli cells. This underlines the need for research on seasonal bat species with one reproductive cycle to evaluate the possible existence of numerical changes in the population of Sertoli cells in individual adults.

PARTICIPATION OF APOPTOSIS IN REGULATING MALE GERMINAL CELLS

Apoptosis is an active, genetically-controlled process that results in cell death. Its central mechanism –an evolutionary development found in organisms from worms to mammals– has three main components: Bcl-2 family proteins, caspases belonging to a family of cysteine proteases that cut after aspartic acid residues, and Apaf-1/CED-4 proteins that transmit the signals made up of Bcl-2 family proteins to the caspases [50]. The biochemical activation of these three components is responsible for the morphological changes that cells undergo during apoptosis [51].

The proliferation of spermatogonia and the spontaneous degeneration of spermatocytes often coincide during normal testicular functioning in mammals, and this is where apoptosis plays a central role in controlling the number of testicular germ cells [52]. The spermatogonia are the cells that frequently suffer apoptosis in the testicles, but this process can also affect spermatocytes and spermatids.

Apoptosis is very important for testicular development. Three key factors seem to control the onset of this type of cell death: (i) the individual's age; (ii) the serum concentration of the follicle-stimulating hormone (FSH) and, probably, of the luteinizing hormone (LH), and the concentration of steroid hormones such as testosterone; and, (iii) the phase of the seminiferous epithelium [53 - 55].

Studies of rats have determined that a significant number of Leydig cells suffer apoptosis during the perinatal period, as the reduction of these cells continues until they become scarce after postnatal day 7 [56], and during puberty (16-24 days of age). Neither the Leydig cells nor the Sertoli cells show evidence of DNA fragmentation [55]. In the germinal cells, a massive, but physiological, wave of apoptosis among the spermatogonia occurs during the first round of spermatogenesis but the elimination of spermatogonia is sporadic in individual adults [57, 58].

Another observation from adult rats is that the highest proportion of germinal cells that suffer apoptosis are the A_{2-4} spermatogonia, and that this proportion decreases in primary and secondary spermatocytes and spermatids, respectively [55, 59 - 62]. The A_{1-2} and intermediate spermatogonia rarely degenerate [63].

But excessive apoptosis can cause infertility due to the massive loss of germinal cells [64]. This emphasizes the need to improve our understanding of the participation of apoptosis and, therefore, its modulation during testicular development. Stimulation of apoptosis in germinal cells has a controlled regulation, but research has shown that stimulation can be increased by certain kinds of stress, such as hyperthermia [65] and oxidative stress [64, 66]. Cell elimination by apoptosis is also potentiated by removing hormonal support [67 - 69], and in conditions of cryptorchidism [70 - 73].

APOPTOSIS IN THE TESTICLES OF MAMMALS WITH SEASONAL REPRODUCTION

Seasonal species present a particular case of the regulation of spermatogenesis because the inactivity of the testicles coincides with low concentrations of plasma gonadotrophin hormones (FSH, LH and prolactin) and of testicular steroid hormones. These endocrine events subsequently propitiate the suspension of spermatogenesis as individuals enter a period of reproductive inactivity [26]. A dramatic reduction of 80-90% in testicular mass during the non-reproductive season suggests that apoptosis may play a key role in spermatogenesis [52].

For many species, measuring and monitoring daily changes in the photoperiod translate into physiological signals due to nocturnal secretion of pineal melatonin [15]. In certain rodents, a short photoperiod or long duration of nocturnal melatonin secretion reduces the reproductive function [74]. In bats, the photoperiod may also play a significant role as a proximal signal of reproduction [11]; although it has only have clear experimental evidence of the effect of photoperiod on the males of the species *Antrozous pallidus* [75].

In the Hungarian hamster, *Phodopus sungurus*, the white-footed mouse, *Peromyscus leucopus*, and the European brown hare, *Lepus europaeus* –all recognized as photosensitive species– an increase in apoptosis is related to the testicular regression that results from exposure to a short photoperiod regime [76 - 78]. In contrast, testicular recrudescence, which is preceded by the release of FSH, LH and, concomitantly, an increase in serum PRL and testosterone [79, 80], is independent of photoperiod augment. Observations show that these endocrine changes induce active cellular regeneration in the testicles, including the proliferation of spermatogonia and angiogenesis [80, 81]. In photo-inhibited white-footed mice, a reduction in testicular apoptosis preceded the spontaneous recrudescence of the testicles after 26 weeks of exposure to the reduced photoperiod [82].

However, contradictory results are reported for the role of apoptosis in testicular regression in the roe deer, another species with seasonal reproduction, in which an inverse relation exists between proliferation and apoptosis, thus suggesting that apoptosis promotes testicular involution [15]. However, more recent work indicates that apoptosis is not the cause of testicular involution in this species [91]. Because of these contradictory results, the role of these two cell processes (apoptosis, spermatogenesis) has been demonstrated in another mammal that reproduces seasonally: the Iberian mole (*Talpa occidentalis*), the only animal described as a true case of hermaphroditism [52, 83]. The females of this and other mole species have two bilateral ovotestes instead of normal ovaries [34, 35,

84]. Moles in the southern Iberian Peninsula reproduce in winter, when males have enlarged testicles and females large uteri with intensified ovarian activity. In summer, the testicles shrink to a quarter of their wintertime mass and ovogenesis is suspended in females, allowing an increase in serum testosterone levels and an increase in the testicular part of the ovotestes [37]. Studies found that apoptosis varies in mole testicles depending on the season of the year, and that this affects, especially, the cells in late zygotene and pachytene during the period of sexual inactivity, but does not differentially affect the number of Sertoli cells. More interesting still is the fact that apoptosis is not responsible for the massive depletion of germinal cells that occurs during testicular regression in moles. Also, a wave of spermatogonia proliferation appears to restore the number of spermatogonia lost during the period of testicular inactivity [52].

An electronic microscopical study of the testes of *Miniopterus inflatus*, a vespertilionid bat from the cold zone of Kenya, carried out during just the initial phase of winter when individuals are in the resting sexual condition, provided qualitative evidence suggesting that apoptosis occurs in both compartments of the testicle (tubular and interstitial). This has led researchers to assume that apoptosis participates in testicular regression in this species [85], but cellular change in the vascular tissues is not considered in this study. However, no studies of other bat species have evaluated degeneration and cell death throughout the testicular cycle.

TESTICULAR CYCLE OF *CORYNORHINUS MEXICANUS* BAT

Corynorhinus mexicanus is a scrotal especies [29]. As part of the postnatal development of males, the presence of the testicles inside the inguinal canal has been demonstrated; the complete testicular descent occurs before puberty, then the testes remain permanently in inguinal position during the annual cycle in adults. The reproductive inactivity of sexually mature specimens, is characterized by the presence of few spermatogonia differentiated by their voluminous form and ovoid nucleus as the only type of germ cell, delimited by lot of column-shaped Sertoli cells with the nucleus in a higher position in the seminiferous epithelium. This stage of the *C. mexicanus* testis is histologically similar to that of the immature bats.

Testicular recrudescence of *C. mexicanus* occurs in May, is characterized by an increase in the diameter of the seminiferous tubules, and in the area of the seminiferous epithelium, which consists of at least three layers and show evidence of mitotic activity. In July, the presence of primary spermatocytes and round spermatids is evident and indicative of the resumption of the stage of the meiotic stage of spermatogenesis [86]. The maximum weight and size of the gonads occurs in August, coincident with full spermatogenesis and the testes mass is approximately 40 times greater than in the period of testicular inactivity [29]. Involution occurs suddenly between September and October; in November, the testes present their lowest values for weight and size.

Because *C. mexicanus*, as in the case of all known plecotine species, there is no histomorphometric analysis focused on determining the possible existence of changes in the organization and the proportion of blood vessels irrigating the testicle, in relation to seasonal drastic changes in the size and in the testicular function, so it is not possible to determine whether there is also a development cycle and involution of testicular vasculature in these species, something that is interesting, considering that in these bats exist a temporal asynchrony in male sexual functions. Particularly, in *C. mexicanus*, a histological study allowed to recognize the presence of an organization of such vasculature similar to what occurs in other mammals, where the testicular artery penetrates the tunica albuginea by the ventral-lateral face of the testis in a cranio-caudal direction of the organ, and internally it branches to providing irrigation that supply the testicular parenchyma with nutrients, hormones, and various blood components necessary for the performance of its functions (spermatogenesis and steroidogenesis).

Fig (**1**), allow recognize some of these vascular elements of the testis and some of their changes in appearance, at different stages of the reproductive cycle of *C. mexicanus*, allowing propose the possibility that in these bat species with a drastic seasonal changes in the testicular size during the reproductive cycle, it present a phenomenon of angiogenesis in the adult testis, which makes them excellent models to study this interesting phenomenon of the development and its control.

However, in general, in the literature there are few reports focused on describing the possible changes in the organization of testicular vasculature in relation to reproductive activity in bats. One of these studies, conducted in *Sturnira lilium*, showed that no changes occurs in the proportion of blood vessels in the intertubular compartment of the testis, considering it is a tropical species with continuous reproduction [87]. In a different way, in *Molossus molossus*, other tropical bat species but with seasonal reproductive pattern, it has been recognized the presence of changes in the proportion of blood vessels in association with the seasonal testicular activity [88]. The anterior is an interesting aspect to take into account, for considering the study of changes in the blood vessel system of the testes in temperate seasonal species in which drastic testicular morphologic changes occur associated to the annual cycle.

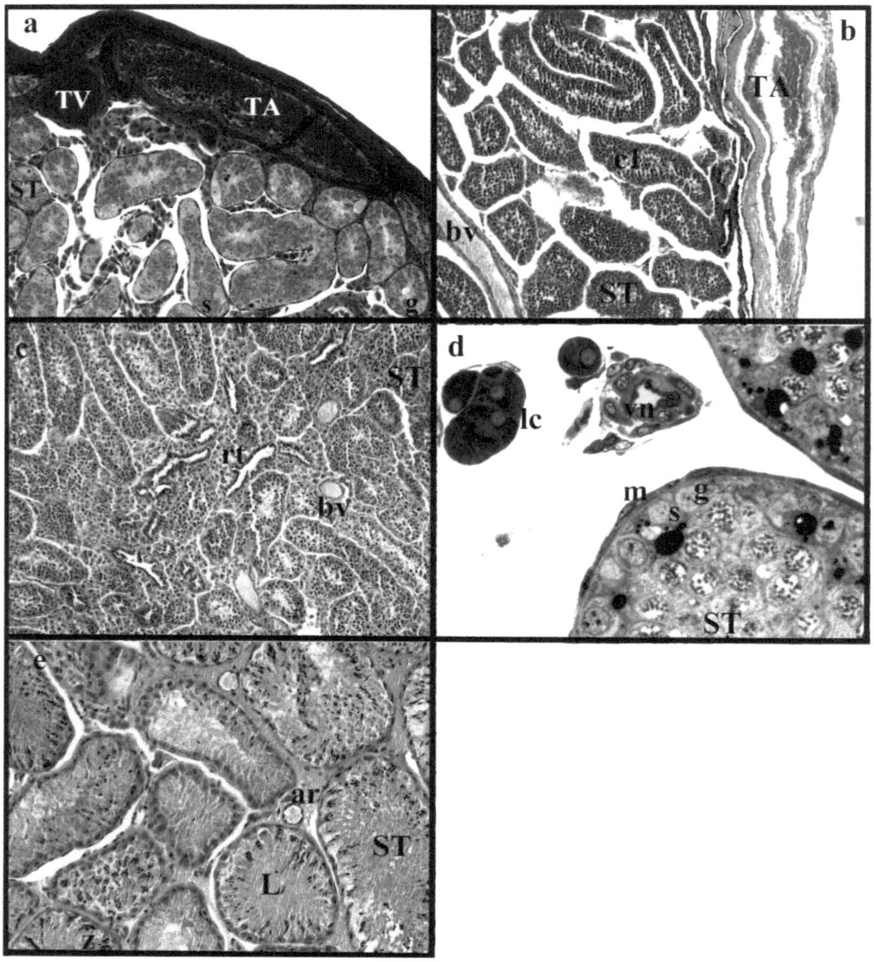

Fig. (1). Photomicrographs of histological sections from the testes of bat specimens captured at key times of the annual cycle, in order to identify the main changes in the microstructural organization of the testicular tissues and the vascular elements associated, trough the testicular cycle.

Abbreviations: ar, arteriole; c1, primary spermatocyte; g, spermatogonia; L, lumen of the seminiferous tubule; lc, Leydig cell; m, perytubular myoid cell; s, Sertoli cell; ST, seminiferous tubule; TA, testicular artery; TC, testicular capsula; TV, testicular vein; ST, seminiferous tubule; bv, blood vessel; rt, rete testis; vn, venula; z, spermatozoa.

a.*February:* inactive stage of spermatogenesis. The seminiferous tubules show no evidence of spermatogenesis. The seminiferous epithelium has some spermatogonia, and most of the Sertoli cells. The blood vessels visible include the artery (the largest) and testicular veins that have penetrated into the interior of the testicle, but are barely sub-capsular.

b.*June:* early stage of spermatogenesis. Seminiferous tubules with epithelium characteristic of mitotic and probable meiotic activity due to a marked increase in the number of germinal cells (spermatogonia and primary spermatocytes) and the presence of many with intensely-stained nuclei. Blood vessels: the testicular artery appears just inside the testicular capsule (extra-testicular vessel) with a branch internalized towards the stroma.

c.*June:* early stage of spermatogenesis. This is the same stage as in (b). Blood vessels: three intra-testicular blood vessels are visible, located in the zone near the rete testis where the sperm cells will congregate at a future time. The thinness of the tunic of these vessels indicates that they are arterioles.
d.*August:* stage of full spermatogenesis. The seminiferous tubules show clear evidence of meiotic activity due to the characteristics of the nuclear chromatin of the cells located in the internal layers of the seminiferous epithelium. Blood vessels: an inter-tubular venula (intra-testicular vessel) with two cell layers is visible.
e.*September:* initial stage of testicular involution. The seminiferous tubules are still enlarged, but there is no evidence of cell proliferation, and the final germinal cells are clearly of the spermatid type in differentiation. Spermatogenesis has ended. Blood vessels: two intra-testicular vessels of the arteriole type are visible immersed in interstitial tissue.
Histological Techniques: for (**a**) and (**d**), fixing with Karnouski; inclusion in epon; toluidine blue staining; section thickness, 1 μ; magnification, 100 x (a) and 1000 x (d).
(**b**), (**c**) and (**e**), Fixing with formalin; inclusion in paraplast plus; H-E staining; section thickness, 8 μ; magnification, 100 x.

APOPTOSIS IN THE TESTICULAR BLOOD VESSELS OF *CORYNORHINUS MEXICANUS*

As it has been described for several scrotal mammals that have been studied, the external location of the testicles requires particular anatomical thermoregulation systems. These consist of a complex vascular testicular network of arteries and veins in the spermatic cord called the pampiniform plexus [89]. In most species, the temperature of arterial blood decreases by 2 to 4ºC before entering the testicles. Two groups of additional muscles –the tunica dartos and cremaster– play an important role in thermoregulation. The first covers the interior of the scrotum and controls its proximity to the testicles. Under low temperature conditions this muscle contracts, pulling the scrotal sac closer to the testicles to provide greater insulation, but as temperatures warm up it relaxes, allowing the scrotum to return to a distal position. The cremaster muscle is located in the spermatic cord. When it contracts during cold periods it pulls the scrotal and testicular sac closer to the base of the body. During warmer climatic conditions it relaxes and allows the testicles to return to their normal position. Both muscles have an abundant supply of nerve fibers that respond to the temperature sensors of the central nervous system.

The blood and vascular lymphatic systems penetrate into the reproductive organs and tissues to supply cells with nutrients and oxygen, in preparation for the circulation of fluids and diverse signaling molecules.

During embryonic development the gonads emerge through the thickening of the mesodermal cells that cover the cavity of the body laterally through the dorsal mesentery, the coelomic epithelium. This thickening is called the genital crest [90, 91]. The migration of primordial germ cells towards the crests depends on two factors: 1) transitory interaction between fibronectin molecules and receptors of the corresponding primordial germ cells, presumably integrins; and, 2) the release of chemoattractant factors by the genital crests [92]. In both sexes, these crests are transformed into undifferentiated gonads.

The number of primordial germ cells in mice embryos is originally below one hundred, but during migration these cells undergo repeated mitotic divisions that increase their number to over 5,000 [90]. Later, the primordial germ cells migrate towards the primitive sexual cords in the core of the gonad, which forms out of the genital crests. The primordial germ cells produce the spermatogonia and, afterwards, the sperm cells [90]. No studies about the embryonic development of the testes have been made by *C. mexicanus* bat.

The emergence of the arterial vascular system (vasculogenesis) is one of the first events in embryogenesis. During early embryonic development, the mesodermal cells differentiate into hemangioblasts, the progenitors of both hematopoietic and endothelial cells, which then form blood vessels. During additional differentiation, the hemangioblasts produce angioblasts, and the aggregation of these elements results in the formation of blood islands. When these islands fuse, they form the primary blood vascular plexus, which is made up of fine capillaries formed by endothelial cells. It is in this stage that the capillaries acquire their character as arteries or veins, indicating that cellular specificity is genetically-programmed [93]. Vasculogenesis is completed together with the formation of the vascular plexus. All later transformations of the vascular network occur during angiogenesis, when new blood vessels form out of existing ones.

In the gonad of male mammals, vasculogenesis is also one of the first characteristics of the organization of the testicular cords, which makes it possible to distinguish morphologically between a testicle and an ovary [3, 94]. During testicular morphogenesis in mice, endothelial cells begin to migrate from the mesonephros at the same time as the organization of the testicular cord begins [5, 85]. Combes *et al.* [95], demonstrated that the migration of endothelial cells and the formation of the testicular cord are actually related and interdependent processes, given that at 11.5 days post-coitus the endothelial mesonephric cells begin to migrate through the gonad towards the coelomic domain due to the accumulation of chemotactic signals in the testicles [96].

The migration of the endothelial mesonephric cells sub-divides the gonad into approximately ten avascular regions where the testicular cords form. At the same time, the vascularized interstitial compartments of the testicle are defined. When the migration of the endothelial cells reaches the testicular coelomic surface they congregate in the coelomic domain where they form the most important testicular artery; the one that runs along the anti-mesonephric border [5, 30]. Thus, the migration of endothelial cells plays a fundamental role in directing the two main processes of early testicular development: the formation of the testicular cord and testicular vascularization [96].

During angiogenesis, the primary vascular plexus expands significantly through capillary ramification until it is transformed into a highly-organized vascular network. Local destruction of the wall of the pre-existing blood vessels begins, leading to the activation of the proliferation of endothelial cells and their migration. The endothelial cells assemble in tubular structures around which the walls of the new blood vessel form. The walls of the finest capillaries have only one layer of cells (pericytes), while those of the arteries and veins are made up of various cell layers of smooth muscle. Pericytes are cells of mesenchymal origin, but their ontogeny is not yet well understood. They comprise a heterogeneous population of cells capable of differentiating into different types of mesenchymal cells, including smooth muscle cells, fibroblasts and osteoblasts.

The strict regulation of angiogenesis and its adequate functioning are very important for organisms, since both the excessive formation and the insufficient development of blood vessels can cause severe diseases [97]. Studies of the formation and involution of testicular blood vessels in seasonal reproductive bat species may thus shed new light on understanding the control and development of angiogenesis in mammals.

The vascular endothelial cells form the interior lining of all blood vessels. During both normal development and pathological conditions, the formation of new vessels and the regression of pre-existing ones depend on the equilibrium between the proliferation and apoptosis of endothelial cells. The rotation of endothelial cells in the mature blood vessels is also controlled by these strongly-regulated phenomena. The physiological remodeling of the blood vessels before and after birth has been shown to result from an equilibrium between apoptosis and cell proliferation, where the development of the programmed regression of capillaries is triggered by the macrophage-mediated apoptosis of endothelial cells [98].

The cells that die are projected towards the lumen of the capillary where they may temporarily or permanently block blood flow. When blood flow ceases, a secondary synchronic apoptosis of vascular endothelial cells occurs that leads to capillary regression [99]. This has generated the hypothesis that during development the selection of capillary beds is controlled by blood flow and depends on the survival or apoptosis of endothelial cells. Cho *et al.* [100], has demonstrated that apoptosis contributes significantly to postpartum arterial remodeling, and that the changes in the rate of cell death can, on their own, be sufficient to induce profound changes in the mass of the vessel walls.

Using the *C. mexicanus* bat as a study model of the seasonal activity of the testicle, the presence of TUNEL-positive cells in the endothelium of extra- and intratesticular blood vessels with a different pattern of marking indicative of apoptosis was observed (Fig. **2**). This information may be considered as the first evidence of involvement of the apoptosis in developing testicular vasculature in a seasonal chiropteran with drastic morphological changes of the testes, although more comparative studies are needed, mainly one between the different stages of testicular cycle which can provide information of the existence of angiogenesis associated with the testicular cycle in these species. Moreover, this differential marking pattern for apoptosis in the endothelium of extra- *vs* intratesticular blood vessels, allows consider a cellular turnover dynamic of the inner layer of blood vessels as part of a maintenance of the vasculature during full spermatogenic stage (Fig. **2**) to provide adequate supply of nutritional and endocrine components that are required to sustain the single annual period of spermatogenesis in these species.

The present study was performed particularly during the stage of complete spermatogenesis of the *C. mexicanus*, so a comparative analysis of the participation of apoptosis in the cells that make up the layers of the different blood vessels of the testis, both intra- and extra-testicular, with which it can be determined whether there is a process of angiogenesis associated with the increase in testicular size and this in turn with the single long period of spermatogenesis (about 5 and a half months in total) that is characteristic of this species, pattern that is similar to other species of seasonal vespertilionids that have asynchrony of their male sexual functions.

As can be seen in Fig. (**2**), there is an important number of cells positively labeled for apoptosis forming part of the outer layers of the extratesticular blood vessels, particularly the testicular artery (Fig. **2a**, **b**), indicating an intense cellular turnover in the vascular tissues during the full spermatogenesis. By contrast, in intratesticular blood vessels of small caliber, such as arterioles, venules and capillaries (Fig. **2c**, **d**), the marking pattern shows apoptotic cells in the lining of the blood vessel, like what happens during angiogenesis that is characteristic of embryonic and early postnatal development of the mouse. In fact, intra-testicular blood vessels show the desquamation of endothelial apoptotic cells that are released into the lumen.

Importantly for seasonal bats of temperate and cold zones of the world, among which are all known plecotinids, especially *C. mexicanus*, when spermatogenic activity is full the testicle reaches its peak in size and mass (up to about 40 times) greater than at the stage of reproductive inactivity [29, 86], and they have clearly shaped to the basal and adluminal compartments of the seminiferous tubule, settling a particular microenvironment that enables the development of the spermatogenesis in connection with the blood-testis barrier structural and functionally well conformed. These particular aspects of the development of structural and functional organization of the testicle in seasonal bats are needed to recognize those details that in set, lead to a successful reproduction of these species in contrasting environments between the different seasons of the annual cycle, and that probably have led to the perpetuation of the species and the maintenance of their populations throughout the time.

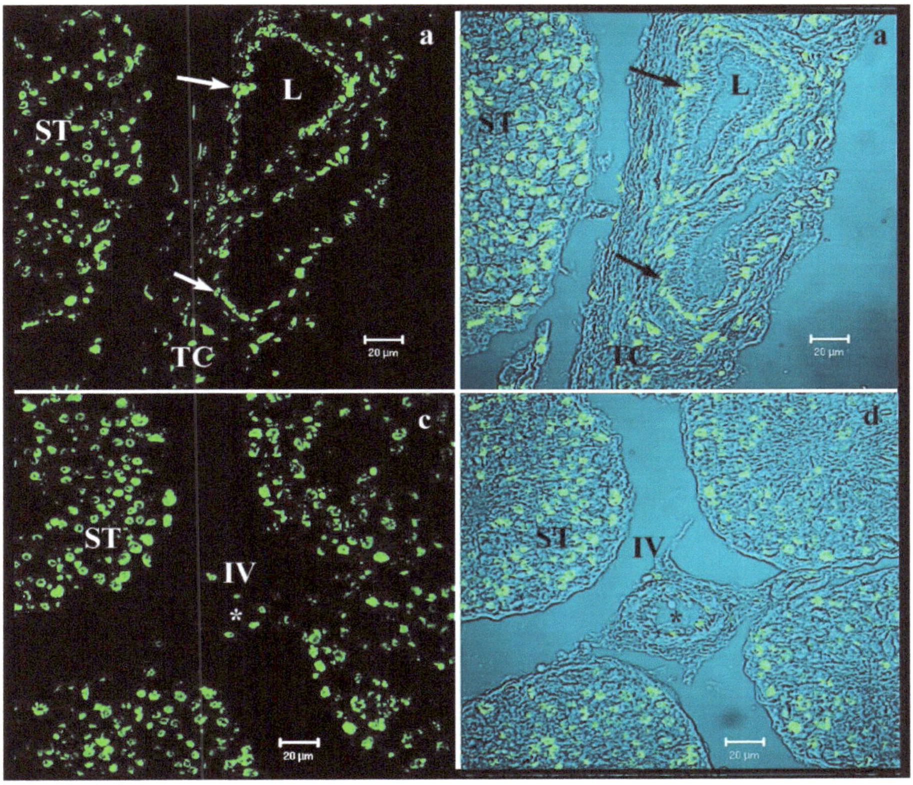

Fig. (2). Photomicrographs of testicular tissue sections TUNEL processed, of *Corynorhinus mexicanus* bat specimens caugth during the full spermatogenic activity (August). Images were obtained from the epifluorescence (**a**, **c**) and Nomarsky contrast (**b**, **d**) tools of the confocal scan lasser microscope. Lines of apoptotic cells (arrows) at external layer are located in extratesticular blood vessels (testicular artery) crossing the capsula (a, b); wereas some apoptotic endothelial cells (asterisk) are identified at intratesticular, intertubular arteriola (**c**, **d**).
IV = intratesticular blood vessel, L = lumen of the testicular artery, ST = seminiferous tubule, TC = testicular capsula.

CONFLICT OF INTEREST

The authors declare no conflict of interest, financial or otherwise.

ACKNOWLEDGEMENTS

Thanks to Dr. Ricardo Lopez Wilchis and Dr. Adolfo Rosado García for their support by providing comments and criticisms to the research that led to a better understanding of the results obtained in the study.

REFERENCES

[1] Hayssen V, Tienhoven AV, Tienhoven AV. Asdell's patterns of mammalian reproduction A compendium of species-specific data. New York: Cornell University Press 1993.

[2] Krutzsch PH. Male reproductive patterns in nonhibernating bats. J Reprod Fertil 1979; 56(1): 333-44. [http://dx.doi.org/10.1530/jrf.0.0560333] [PMID: 381651]

[3] Krutzsch PH. Anatomy, physiology and cyclicity of the male reproductive tract. In: Crichton EG, Krutzsch PH, Eds. Reproductive Biology of Bats. Academic Press 2000; pp. 91-155. [http://dx.doi.org/10.1016/B978-012195670-7/50005-9]

[4] Tamsitt JR, Valdivieso D. Reproductive cycle of the big fruit-eating bat, artibeus lituratus olfers. Nature 1963; 198(4875): 104-4. [http://dx.doi.org/10.1038/198104a0]

[5] Racey PA, Entwistle AC. Life-history and reproductive strategies of bats. In: Crichton EG, Krutzsch PH, Eds. Reproductive Biology of Bats. Academic Press 2000; pp. 364-414. [http://dx.doi.org/10.1016/B978-012195670-7/50010-2]

[6] Wimsatt W. Some problems of reproduction in relation to hibernation in bats. Bulletin of the Museum of Comparative Zoology Harvard 1960; 124: 249-67.

[7] Wimsatt WA. Some interrelations of reproduction and hibernation in mammals. Symp Soc Exp Biol 1969; 23: 511-49. [PMID: 5367180]

[8] Entwistle A, Racey P, Speakman J. The reproductive cycle and determination of sexual maturity in male brown long-eared bats, Plecotus auritus (Chiroptera: Vespertilionidae). J Zool 1998; 244(01): 63-70. [http://dx.doi.org/10.1111/j.1469-7998.1998.tb00007.x]

[9] Bogdanowicz W, Kasper S, Owen RD. Phylogeny of plecotine bats: reevaluation of morphological and chromosomal data. J Mammal 1998; 79(1): 78-90. [http://dx.doi.org/10.2307/1382843]

[10] Racey PA. Ecology of Bat Reproduction. In: Kunz TH, Ed. Ecology of Bats. Boston, MA: Springer US 1982; pp. 57-104. [http://dx.doi.org/10.1007/978-1-4613-3421-7_2]

[11] Heideman PD. Environmental regulation of reproduction. In: Crichton EG, Krutzsch PH, Eds. Reproductive biology of bats. Academic Press 2000; pp. 469-500. [http://dx.doi.org/10.1016/B978-012195670-7/50012-6]

[12] Alexandre-Pires G, Mateus L, Martins C, Ferreira-Dias G. Seasonal changes in testes vascularisation in the domestic cat (Felis domesticus): evaluation of microvasculature, angiogenic activity, and endothelial cell expression. Anatomy research international 2012.

[13] Malpaux B. Seasonal regulation of reproduction in mammals. In: Knobil E, Neill JD, Eds. Knobil and Neills Physiology of Reproduction. Gulf Professional Publishing 2006; pp. 2231-82. [http://dx.doi.org/10.1016/B978-012515400-0/50046-4]

[14] Folch J. Influence of age, photoperiodism and temperature on semen production of rams. In: Courot M, Ed. The male in farm animal reproduction. Springer Netherlands 1984; pp. 141-60.

[15] Bronson FH, Heideman PD. Seasonal regulation of reproduction in mammals. In: Knobil E, Neill JD, Eds. The physiology of reproduction. 1994; pp. 541-84.

[16] Morgentaler A, Stahl BC, Yin Y. Testis and temperature: an historical, clinical, and research perspective. J Androl 1999; 20(2): 189-95.
[PMID: 10232653]

[17] León-Galván MA, Fonseca T, López-Wilchis R, Rosado A. Prolonged storage of spermatozoa in the genital tract of female Mexican big-eared bats (Corynorhinus mexicanus): the role of lipid peroxidation. Can J Zool 1999; 77(1): 7-12.
[http://dx.doi.org/10.1139/z98-197]

[18] Arenas-Ríos E, Rosado García A, Cortés-Barberena E, *et al.* Reactive oxygen species production and antioxidant enzyme activity during epididymal sperm maturation in Corynorhinus mexicanus bats. Reprod Biol 2016; 16(1): 78-86.
[http://dx.doi.org/10.1016/j.repbio.2016.01.001] [PMID: 26952757]

[19] Bicego KC, Barros RC, Branco LG. Physiology of temperature regulation: comparative aspects. Comp Biochem Physiol A Mol Integr Physiol 2007; 147(3): 616-39.
[http://dx.doi.org/10 1016/j.cbpa.2006.06.032] [PMID: 16950637]

[20] Mital P, Hinton BT, Dufour JM. The blood-testis and blood-epididymis barriers are more than just their tight junctions. Biol Reprod 2011; 84(5): 851-8.
[http://dx.doi.org/10.1095/biolreprod.110.087452] [PMID: 21209417]

[21] Niessen CM. Tight junctions/adherens junctions: basic structure and function. J Invest Dermatol 2007; 127(11): 2525-32.
[http://dx.doi.org/10.1038/sj.jid.5700865] [PMID: 17934504]

[22] Wolburg H, Lippoldt A. Tight junctions of the blood-brain barrier: development, composition and regulation. Vascul Pharmacol 2002; 38(6): 323-37.
[http://dx.doi.org/10.1016/S1537-1891(02)00200-8] [PMID: 12529927]

[23] Su L, Mruk DD, Cheng CY. Drug transporters, the blood-testis barrier, and spermatogenesis. J Endocrinol 2011; 208(3): 207-23.
[PMID: 21134990]

[24] McLachlan RI, O'Donnell L, Meachem SJ, *et al.* Identification of specific sites of hormonal regulation in spermatogenesis in rats, monkeys, and man. Recent Prog Horm Res 2002; 57: 149-79.
[http://dx.doi.org/10.1210/rp.57.1.149] [PMID: 12017541]

[25] Yoshida S, Sukeno M, Nabeshima Y. A vasculature-associated niche for undifferentiated spermatogonia in the mouse testis. Science 2007; 317(5845): 1722-6.
[http://dx.doi.org/10.1126/science.1144885] [PMID: 17823316]

[26] Bronson FH. Mammalian Reproductive Biology. University of Chicago Press 1989.

[27] Krishna A, Singh K. The relationship between testicular activity, accessory sex glands, and circulating steroid concentration during the reproductive cycle in a male Indian vespertilionid bat, Scotophilus heathi. Can J Zool 1997; 75(7): 1042-50.
[http://dx.doi.org/10.1139/z97-125]

[28] Martin L, Bernard RT. Endocrine regulation of reproduction in bats: the role of circulating gonadal hormones. In: Crichton EG, Krutzsch PH, Eds. Reproductive biology of bats. London: Academic Press 2000; pp. 27-64.
[http://dx.doi.org/10.1016/B978-012195670-7/50003-5]

[29] León-Galván MA, López-Wilchis R, Hernández-Pérez O, *et al.* Male reproductive cycle of Mexican big-eared bats, *Corynorhinus mexicanus* (Chiroptera: Vespertilionidae). Southwest Nat 2005; 50(4): 453-60.
[http://dx.doi.org/10.1894/0038-4909(2005)050[0453:MRCOMB]2.0.CO;2]

[30] Speakman JR, Racey PA. The influence of body condition on sexual development of male Brown long-eared bats (Plecotus auritus) in the wild. J Zool 1986; 210(4): 515-25. [http://dx.doi.org/10.1111/j.1469-7998.1986.tb03653.x]

[31] Jolly SE, Blackshaw AW. Prolonged epididymal sperm storage, and the temporal dissociation of testicular and accessory gland activity in the common sheath-tail bat, Taphozous georgianus, of tropical Australia. J Reprod Fertil 1987; 81(1): 205-11. [http://dx.doi.org/10.1530/jrf.0.0810205] [PMID: 3668950]

[32] Kurta A, Kunz TH. Roosting metabolic rate and body temperature of male little brown bats (Myotis lucifugus) in summer. J Mammal 1988; 69(3): 645-51. [http://dx.doi.org/10.2307/1381365]

[33] Pearson OP, Koford MR, Pearson AK. Reproduction of the lump-nosed bat (Corynorhinus rafinesquei) in California. J Mammal 1952; 33(3): 273-320. [http://dx.doi.org/10.2307/1375769]

[34] Sinha Hikim AP, Bartke A, Russell LD. Morphometric studies on hamster testes in gonadally active and inactive states: light microscope findings. Biol Reprod 1988; 39(5): 1225-37. [http://dx.doi.org/10.1095/biolreprod39.5.1225] [PMID: 3219392]

[35] Hombach-Klonisch S, Schön J, Kehlen A, *et al.* Seasonal expression of INSL3 and Lgr8/Insl3 receptor transcripts indicates variable differentiation of Leydig cells in the roe deer testis. Biol Reprod 2004; 71(4): 1079-87. [http://dx.doi.org/10.1095/biolreprod.103.024752] [PMID: 15151926]

[36] Hardy MP, Mendis-Handagama SM, Zirkin BR, Ewing LL. Photoperiodic variation of Leydig cell numbers in the testis of the golden hamster: a possible mechanism for their renewal during recrudescence. J Exp Zool 1987; 244(2): 269-76. [http://dx.doi.org/10.1002/jez.1402440211] [PMID: 3430123]

[37] Johnson L, Tatum ME. Temporal appearance of seasonal changes in numbers of Sertoli cells, Leydig cells, and germ cells in stallions. Biol Reprod 1989; 40(5): 994-9. [http://dx.doi.org/10.1095/biolreprod40.5.994] [PMID: 2765623]

[38] Hochereau-de Reviers MT, Lincoln GA. Seasonal variation in the histology of the testis of the red deer, Cervus elaphus. J Reprod Fertil 1978; 54(2): 209-13. [http://dx.doi.org/10.1530/jrf.0.0540209] [PMID: 722668]

[39] Komatsu T, Yamamoto Y, Atoji Y, *et al.* Seasonal changes in subcellular structures of Leydig and Sertoli cells in the Japanese black bear, Ursus thibetanus japonicus. Arch Histol Cytol 1997; 60(3): 225-34. [http://dx.doi.org/10.1679/aohc.60.225] [PMID: 9376170]

[40] Sharpe RM. Regulation of spermatogenesis. In: Knobil E, Ed. The physiology of reproduction. New York: Raven Press 1994; pp. 1363-434.

[41] Muñoz EM, Fogal T, Dominguez S, *et al.* Ultrastructural and morphometric study of the Sertoli cell of the viscacha (Lagostomus maximus maximus) during the annual reproductive cycle. Anat Rec 2001; 262(2): 176-85. [http://dx.doi.org/10.1002/1097-0185(20010201)262:2<176::AID-AR1022>3.0.CO;2-I] [PMID: 11169912]

[42] Steinberger A, Steinberger E. The Sertoli cells. In: Johnson AD, Gomes WR, Eds. The testis. New York: Academic Press 1977; pp. 371-99.

[43] Nagy F. Cell division kinetics and DNA synthesis in the immature sertoli cells of the rat testis. J Reprod Fertil 1972; 28(3): 389-95. [http://dx.doi.org/10.1530/jrf.0.0280389] [PMID: 5014338]

[44] De Kretser DM, Kerr JB. The cytology of the testis. In: Knobil E, Neill DJ, Eds. The physiology of reproduction. New York: Raven Press 1994; pp. 1177-290.

[45] Nandi S, Banerjee PP, Zirkin BR. Germ cell apoptosis in the testes of Sprague Dawley rats following testosterone withdrawal by ethane 1,2-dimethanesulfonate administration: relationship to Fas? Biol Reprod 1999; 61(1): 70-5.
[http://dx.doi.org/10.1095/biolreprod61.1.70] [PMID: 10377033]

[46] Johnson L, Varner DD, Tatum ME, Scrutchfield WL. Season but not age affects Sertoli cell number in adult stallions. Biol Reprod 1991; 45(3): 404-10.
[http://dx.doi.org/10.1095/biolreprod45.3.404] [PMID: 1782288]

[47] Meachem SJ, Stanton PG, Schlatt S. Follicle-stimulating hormone regulates both Sertoli cell and spermatogonial populations in the adult photoinhibited Djungarian hamster testis. Biol Reprod 2005; 72(5): 1187-93.
[http://dx.doi.org/10.1095/biolreprod.104.039321] [PMID: 15659702]

[48] Hochereau-de Reviers MT, Perreau C, Lincoln GA. Photoperiodic variations of somatic and germ cell populations in the Soay ram testis. J Reprod Fertil 1985; 74(2): 329-34.
[http://dx.doi.org/10.1530/jrf.0.0740329] [PMID: 3930709]

[49] Bielli A, Pedrana G, Gastel MT, *et al.* Influence of grazing management on the seasonal change in testicular morphology in Corriedale rams. Anim Reprod Sci 1999; 56(2): 93-105.
[http://dx.doi.org/10.1016/S0378-4320(99)00027-5] [PMID: 10463397]

[50] Adams JM, Cory S. The Bcl-2 protein family: arbiters of cell survival. Science 1998; 281(5381): 1322-6.
[http://dx.doi.org/10.1126/science.281.5381.1322] [PMID: 9735050]

[51] Thornberry NA, Lazebnik Y. Caspases: enemies within. Science 1998; 281(5381): 1312-6.
[http://dx.doi.org/10.1126/science.281.5381.1312] [PMID: 9721091]

[52] Dadhich RK, Real FM, Zurita F, *et al.* Role of apoptosis and cell proliferation in the testicular dynamics of seasonal breeding mammals: a study in the Iberian mole, Talpa occidentalis. Biol Reprod 2010; 83(1): 83-91.
[http://dx.doi.org/10.1095/biolreprod.109.080135] [PMID: 20357272]

[53] Bartke A. Apoptosis of male germ cells, a generalized or a cell type-specific phenomenon? Endocrinology 1995; 136(1): 3-4.
[http://dx.doi.org/10.1210/endo.136.1.7828545] [PMID: 7828545]

[54] Terranova PF, Taylor CC. Apoptosis (cell death). In: Knobil E, Neill DJ, Eds. Encyclopedia of reproduction. London, UK: Academic Press 1999; pp. 261-73.

[55] Billig H, Furuta I, Rivier C, *et al.* Apoptosis in testis germ cells: developmental changes in gonadotropin dependence and localization to selective tubule stages. Endocrinology 1995; 136(1): 5-12.
[http://dx.doi.org/10.1210/endo.136.1.7828558] [PMID: 7828558]

[56] Yokoi Hiromichi, Yamama Yoshihiro, Tsuruo Yoshihiro, *et al.* Leydig Cells Undergoing Apoptosis in the Perinatal Rat Testis. Acta Histochem Cytochem 1998; 31(4): 355-61.
[http://dx.doi.org/10.1267/ahc.31.355]

[57] Rodriguez I, Ody C, Araki K, *et al.* An early and massive wave of germinal cell apoptosis is required for the development of functional spermatogenesis. EMBO J 1997; 16(9): 2262-70.
[http://dx.doi.org/10.1093/emboj/16.9.2262] [PMID: 9171341]

[58] Blanco-Rodríguez J. A matter of death and life: the significance of germ cell death during spermatogenesis. Int J Androl 1998; 21(5): 236-48.
[http://dx.doi.org/10.1046/j.1365-2605.1998.00133.x] [PMID: 9805237]

[59] Allan DJ, Harmon BV, Roberts SA. Spermatogonial apoptosis has three morphologically recognizable phases and shows no circadian rhythm during normal spermatogenesis in the rat. Cell Prolif 1992; 25(3): 241-50.
[http://dx.doi.org/10.1111/j.1365-2184.1992.tb01399.x] [PMID: 1596537]

[60] Kerr JB. Spontaneous degeneration of germ cells in normal rat testis: assessment of cell types and frequency during the spermatogenic cycle. J Reprod Fertil 1992; 95(3): 825-30. [http://dx.doi.org/10.1530/jrf.0.0950825] [PMID: 1404097]

[61] Blanco-Rodríguez J, Martínez-García C. Spontaneous germ cell death in the testis of the adult rat takes the form of apoptosis: re-evaluation of cell types that exhibit the ability to die during spermatogenesis. Cell Prolif 1996; 29(1): 13-31. [http://dx.doi.org/10.1111/j.1365-2184.1996.tb00091.x] [PMID: 8603107]

[62] Brinkworth MH, Weinbauer GF, Schlatt S, Nieschlag E. Identification of male germ cells undergoing apoptosis in adult rats. J Reprod Fertil 1995; 105(1): 25-33. [http://dx.doi.org/10.1530/jrf.0.1050025] [PMID: 7490711]

[63] Huckins C. The morphology and kinetics of spermatogonial degeneration in normal adult rats: an analysis using a simplified classification of the germinal epithelium. Anat Rec 1978; 190(4): 905-26. [http://dx.doi.org/10.1002/ar.1091900410] [PMID: 637327]

[64] Fujii J, Iuchi Y, Matsuki S, Ishii T. Cooperative function of antioxidant and redox systems against oxidative stress in male reproductive tissues. Asian J Androl 2003; 5(3): 231-42. [PMID: 12937808]

[65] Miura M, Sasagawa I, Suzuki Y, *et al.* Apoptosis and expression of apoptosis-related genes in the mouse testis following heat exposure. Fertil Steril 2002; 77(4): 787-93. [http://dx.doi.org/10.1016/S0015-0282(01)03255-1] [PMID: 11937135]

[66] Ikeda M, Kodama H, Fukuda J, *et al.* Role of radical oxygen species in rat testicular germ cell apoptosis induced by heat stress. Biol Reprod 1999; 61(2): 393-9. [http://dx.doi.org/10.1095/biolreprod61.2.393] [PMID: 10411517]

[67] Hikim AP, Wang C, Leung A, Swerdloff RS. Involvement of apoptosis in the induction of germ cell degeneration in adult rats after gonadotropin-releasing hormone antagonist treatment. Endocrinology 1995; 136(6): 2770-5. [http://dx.doi.org/10.1210/endo.136.6.7750502] [PMID: 7750502]

[68] Sinha Hikim AP, Rajavashisth TB, Sinha Hikim I, *et al.* Significance of apoptosis in the temporal and stage-specific loss of germ cells in the adult rat after gonadotropin deprivation. Biol Reprod 1997; 57(5): 1193-201. [http://dx.doi.org/10.1095/biolreprod57.5.1193] [PMID: 9369187]

[69] Woolveridge I, de Boer-Brouwer M, Taylor MF, *et al.* Apoptosis in the rat spermatogenic epithelium following androgen withdrawal: changes in apoptosis-related genes. Biol Reprod 1999; 60(2): 461-70. [http://dx.doi.org/10.1095/biolreprod60.2.461] [PMID: 9916015]

[70] Ohta Y, Nishikawa A, Fukazawa Y, *et al.* Apoptosis in adult mouse testis induced by experimental cryptorchidism. Acta Anat (Basel) 1996; 157(3): 195-204. [http://dx.doi.org/10.1159/000147881] [PMID: 9226038]

[71] Barqawi A, Trummer H, Meacham R. Effect of prolonged cryptorchidism on germ cell apoptosis and testicular sperm count. Asian J Androl 2004; 6(1): 47-51. [PMID: 15064834]

[72] Chaki SP, Misro MM, Ghosh D, *et al.* Apoptosis and cell removal in the cryptorchid rat testis. Apoptosis 2005; 10(2): 395-405. [http://dx.doi.org/10.1007/s10495-005-0813-7] [PMID: 15843900]

[73] Ofordeme KG, Aslan AR, Nazir TM, *et al.* Apoptosis and proliferation in human undescended testes. BJU Int 2005; 96(4): 634-8. [http://dx.doi.org/10.1111/j.1464-410X.2005.05698.x] [PMID: 16104924]

[74] Blank JL, Freeman DA. Differential reproductive response to short photoperiod in deer mice: role of melatonin. J Comp Physiol A Neuroethol Sens Neural Behav Physiol 1991; 169(4): 501-6. [http://dx.doi.org/10.1007/BF00197662] [PMID: 1779421]

[75] Beasley LJ, Zucker I. Photoperiod influences the annual reproductive cycle of the male pallid bat (Antrozous pallidus). J Reprod Fertil 1984; 70(2): 567-73. [http://dx.doi.org/10.1530/jrf.0.0700567] [PMID: 6699816]

[76] Furuta I, Porkka-Heiskanen T, Scarbrough K, *et al.* Photoperiod regulates testis cell apoptosis in Djungarian hamsters. Biol Reprod 1994; 51(6): 1315-21. [http://dx.doi.org/10.1095/biolreprod51.6.1315] [PMID: 7888511]

[77] Young KA, Zirkin BR, Nelson RJ. Short photoperiods evoke testicular apoptosis in white-footed mice (Peromyscus leucopus). Endocrinology 1999; 140(7): 3133-9. [http://dx.doi.org/10 1210/endo.140.7.6870] [PMID: 10385406]

[78] Young KA, Zirkin BR, Nelson RJ. Testicular regression in response to food restriction and short photoperiod in white-footed mice (Peromyscus leucopus) is mediated by apoptosis. Biol Reprod 2000; 62(2): 347-54. [http://dx.doi.org/10.1095/biolreprod62.2.347] [PMID: 10642572]

[79] Bartke A, Goldman BD, Bex FJ, *et al.* Effects of prolactin on testicular regression and recrudescence in the golden hamster. Endocrinology 1980; 106(1): 167-72. [http://dx.doi.org/10.1210/endo-106-1-167] [PMID: 6243094]

[80] Berkowitz AS, Heindel JJ. Testicular recrudescence in the golden hamster (Mesocricetus auratus): a possible model of sexual maturation. Endocrinology 1984; 114(3): 855-60. [http://dx.doi.org/10.1210/endo-114-3-855] [PMID: 6321143]

[81] Mayerhofer A, Sinha Hikim AP, Bartke A, Russell LD. Changes in the testicular microvasculature during photoperiod-related seasonal transition from reproductive quiescence to reproductive activity in the adult golden hamster. Anat Rec 1989; 224(4): 495-507. [http://dx.doi.org/10.1002/ar.1092240407] [PMID: 2476952]

[82] Young KA, Zirkin BR, Nelson RJ. Testicular apoptosis is down-regulated during spontaneous recrudescence in white-footed mice (Peromyscus leucopus). J Biol Rhythms 2001; 16(5): 479-88. [http://dx.doi.org/10.1177/074873001129002150] [PMID: 11669421]

[83] Neaves WB. Changes in testicular leydig cells and in plasma testosterone levels among seasonally breeding rock hyrax. Biol Reprod 1973; 8(4): 451-66. [http://dx.doi.org/10.1093/biolreprod/8.4.451] [PMID: 4710785]

[84] Johnson L, Matt KS, Bartke A, *et al.* Effect of photoperiod on the size of the Leydig cell population and the rate of recruitment of Leydig cells in adult Syrian hamsters. Biol Reprod 1987; 37(3): 727-38. [http://dx.doi.org/10.1095/biolreprod37.3.727] [PMID: 3676416]

[85] Hendrickx AG, Otiang'a-owiti GE, Gachoka JM, Onyango DW. Seasonally dependent testicular apoptosis in the tropical Long-fingered bat (Miniopterus inflatus). Z Saugetierkd 1995; 60(4): 206-14.

[86] León-Galván MA. Participación de la muerte celular por apoptosis durante la espermatogénesis en el murciélago (Corynorhinus mexicanus) (Chiroptera: Vespertilionidae) Doctoral thesis . México, DF: Universidad Autónoma Metropolitana-Iztapalapa 2008.

[87] Morais DB, Barros MS, Freitas MB, *et al.* Histomorphometric characterization of the intertubular compartment in the testes of the bat Sturnira lilium. Anim Reprod Sci 2014; 147(3-4): 180-6. [http://dx.doi.org/10.1016/j.anireprosci.2014.03.008] [PMID: 24793584]

[88] Morais DB, De Oliveira LC, Carmo Cupertino MD, *et al.* Organization and seasonal quantification of the intertubular compartment in the bat Molossus molossus (Pallas, 1776) testis. Microsc Res Tech 2013; 76(1): 94-101. [http://dx.doi.org/10.1002/jemt.22141] [PMID: 23077089]

[89] Garner DL, Hafez ES. Spermatozoa and seminal plasma. In: Hafez ES, Ed. Reproduction in Farm Animals. Philadelphia: Lea and Febiger 1993; pp. 165-87.

[90] Balinsky BI. An introduction to embryology Saunders. Philadelphia 1970.

[91] Setchell BP. Naturally occurring and induced dysfunctions of the testis. In: Setchell BP, Ed. The mammalian testis. London: Paul Eleck 1978; pp. 359-42.

[92] Kierszenbaum AL. Mammalian spermatogenesis in vivo and in vitro: a partnership of spermatogenic and somatic cell lineages. Endocr Rev 1994; 15(1): 116-34. [PMID: 8156936]

[93] Carmeliet P, Collen D. Vascular development and disorders: molecular analysis and pathogenic insights. Kidney Int 1998; 53(6): 1519-49. [http://dx.doi.org/10.1046/j.1523-1755.1998.00936.x] [PMID: 9607184]

[94] Geiser F, Ruf T. Hibernation versus Daily Torpor in Mammals and Birds: Physiological Variables and Classification of Torpor Patterns. Physiol Zool 1995; 68(6): 935-66. [http://dx.doi.org/10.1086/physzool.68.6.30163788]

[95] Combes AN, Wilhelm D, Davidson T, *et al.* Endothelial cell migration directs testis cord formation. Dev Biol 2009; 326(1): 112-20. [http://dx.doi.org/10.1016/j.ydbio.2008.10.040] [PMID: 19041858]

[96] Caruso M, Ferranti F, Corano Scheri K, *et al.* R-spondin 1/dickkopf-1/beta-catenin machinery is involved in testicular embryonic angiogenesis. PLoS One 2015; 10(4): e0124213. [http://dx.doi.org/10.1371/journal.pone.0124213] [PMID: 25910078]

[97] Karamysheva AF. Mechanisms of angiogenesis. Biochemistry (Mosc) 2008; 73(7): 751-62. [http://dx.doi.org/10.1134/S0006297908070031] [PMID: 18707583]

[98] Lang RA, Bishop JM. Macrophages are required for cell death and tissue remodeling in the developing mouse eye. Cell 1993; 74(3): 453-62. [http://dx.doi.org/10.1016/0092-8674(93)80047-I] [PMID: 8348612]

[99] Meeson A, Palmer M, Calfon M, Lang R. A relationship between apoptosis and flow during programmed capillary regression is revealed by vital analysis. Development 1996; 122(12): 3929-38. [PMID: 9012513]

[100] Cho A, Courtman DW, Langille BL. Apoptosis (programmed cell death) in arteries of the neonatal lamb. Circ Res 1995; 76(2): 168-75. [http://dx.doi.org/10.1161/01.RES.76.2.168] [PMID: 7834826]

CHAPTER 5

Epididymal Sperm Maturation in Bats

Edith Arenas-Ríos[1,*], Ahiezer Rodríguez-Tobón[1] and Miguel Ángel León-Galván[2]

[1]*Departamento de Biología de la Reproducción, Universidad Autónoma Metropolitana - Iztapalapa, Ciudad de México, México*

[2] *Departamento de Biología, Universidad Autónoma Metropolitana - Iztapalapa, Ciudad de México, México*

Abstract: After spermatogenesis within the seminiferous tubules of the male gonads, the sperm cells continue their journey through the male reproductive tract, until ejaculation, however, described for the generality of mammals studied, that transit through the epididymis is essential for sperm to acquire the potential to carry out hyperactivation and acrosome reaction within the female genital tract and, in turn, fertilize the egg. The sperm is a transcriptionally silenced cell, so it depends almost entirely on proteins that are acquired or modified during their tour of the epididymal duct. Recently, studies have been conducted using the bat (Corynorhinum mexicanus) as a study model, which has allowed to explore some important details of the process of epididymal sperm maturation, such as the fact that in this species, the epididymal maturation ends in the caudal region, and not as had been established so far in other groups of mammals, another aspect to highlight is the participation of the reactive oxygen species and the different enzymes involved, which have been reported to play a crucial role in the signaling pathways within the maturation of sperm. Also, studies have been conducted on post-translational changes that present the proteins that make sperm and its various organelles, fundamentals for the survival of the cell and its recognition with the ovum.

Keywords: Bat, Chiroptera, Vespertilionidae, Prolonged storage sperm, Sperm, Sperm maturation, Storage sperm.

INTRODUCTION

A discussion of Epididymal Sperm Maturation (ESM) in bats could begin with the reports of what happens during maturation in other mammals, but male bats exhibit great diversity in their reproductive patterns in terms of the timing and frequency of their annual reproductive cycles. There are species with only one cycle per year and some of these present asynchrony between the functions of the

* **Corresponding author Edith Arenas Ríos:** Department of Biology of Reproduction, Autonomous Metropolitan University-Iztapalapa, Ex de San Juan de Dios, 14387 Ciudad de México, CDMX, Mexico; Tel: +55 5804 4923; E-mail: editharenas2000@yahoo.cm.mx

Edith Arenas-Ríos (Ed.)

testicle and epididymis, such that the sperms produced in the former at a certain time of year are stored in the caudal region of the latter for several months. This is particularly interesting when we consider that, on one hand, sperm transit and storage in most mammal species require only some 10-20 days [1]; on the other hand, the ESM process is dependent on androgens produced by the testicle [2]. Epididymal sperm maturation is necessary because the sperms released by the testicle do not acquire progressive motility or the ability to fertilize the oocyte. These capacities develop as they pass through the epididymis, each region of which brings them into direct contact with specific luminal microenvironments. This, in turn, depends –among other phenomena– on gene expression and protein synthesis by the epididymis' epithelial cells, which also participate in maturation [1, 3 - 6]. Another important aspect is how proteins are modified during transit. Most proteins in mature sperms are synthetized in the testicle, but the nucleus of the sperm is highly-condensed and transcriptionally inactive [6 - 8], which means that ESM also depends largely on modifying the proteins that are incorporated, either from the testicle, or during passage through the epididymis. The result of all these changes, which finally give the sperm its fertilizing capacity upon emerging from the epididymis, is called Epididymal Sperm Maturation (ESM).

As this chapter will show, ESM plays a fundamental role in enabling the sperm –after ejaculation– to navigate the female reproductive tract, cross the cumulus cells and pellucid zone, recognize the oocyte and fertilize it [9 - 11]. The chapter begins with a description of the reproductive cycles of temperate zone bats, the group which contains the species that present adaptations distinguish them from most mammals and giving them their peculiar nature. Second, it discusses in detail the specific modifications of ESM in bats generated by these adaptations.

REPRODUCTION IN TEMPERATE ZONE BATS

Male reproductive periodicity in mammals can be analyzed on the basis of different generalized patterns that characterize this group of vertebrates. These can be divided into two broad categories:

The first group includes species that reproduce more than once a year, either because recrudescence and testicular involution occur bimodally, compared to the bimodal polyestrous or continuous reproductive pattern characteristic of females that includes over two annual estrous cycles. In the latter, it is possible that not all males in the population will be reproductively active at the same time.

The second group includes the species in which testicular activity occurs only once a year linked, as in all mammalian reproductive patterns, to the female's monoestrous cycle. This group includes many mammals that inhabit temperate zones.

Male Reproductive Patterns

When dealing with bats, it is important to note that various species have adapted to life in climatically temperate regions (*i.e.*, latitudes above the tropics). As a result, some belong principally to the *Vespertilionidae* and *Rhinolophidae* families, which present recurring annual periods of physiological dormancy during the cold season [12 - 17]. When this condition (torpor) is prolonged, it is called hibernation [18]. But other bats show intermediate thermoregulatory patterns, such as intermittent torpor or daily torpor [18, 19]. In certain cases, bats with this pattern of intermediate dormancy inhabit climatically temperate regions propitiated by the high altitudes at which they live.

One characteristic of almost all non-chiroptera mammals is that the annual periods of hibernation and reproduction do not overlap, at least not significantly, so it is thought that the deep metabolic depression that marks hibernation exerts no direct influence on reproductive activity except –possibly– for delaying its onset [18]. Examples of species that manifest this phenomenon are: *Marmota marmota*, *Erinaceous europaeus*, *Spermophilus citelus* and *Mesocricetus auratus* [20], all typically spring reproducers with seasonal spermatogenesis [21].

In contrast, many bats that exhibit winter dormancy present a singular difference with respect to the aforementioned pattern. In these species, dormancy *directly* overlaps the reproductive period, so torpor markedly affects the reproductive physiology of both sexes [12, 18, 22, 23]. While the females of all these bats have a monoestrous pattern [12, 24], males have just one limited peak of spermatogenesis [12, 22, 25]. A second key difference compared to other hibernating mammals is that the male reproductive pattern of certain bats in the *Vespertilionidae* and *Rhinolophidae* families is that, resumption of spermatogenesis is asynchronous in terms of the activities of the epididymis and the secondary sexual organs, respectively. Asynchrony means that the annual cycles of the testicle, the epididymis and the accessory sex glands, are temporarily out-of-phase [12, 18, 22, 26, 27]. Briefly, spermatogenic activity resumes each spring, shortly after emergence from hibernation, reaches its peak from mid-t--late summer, and decreases rapidly to a condition similar to the pre-pubertal state by late summer or early autumn, when mating begins. Here, sperm production and mating behavior are de-coupled [12, 22, 23, 26, 28, 29].

In marked contrast to testicular activity in those bats, the secondary sex organs (including the tail of the epididymis (sperm storage area) and the accessory sex glands) experience late recrudescence and hypertrophy and become replete with sperm (epididymis) or secretions (glands) during the overlapping intervals of mating and torpor [23]. Later, the animals awake permanently from their lethargic condition and their secondary sex organs –generally– undergo complete regression to produce a condition similar to pre-pubescence in late spring-early summer, when testicular activity resumes [23].

A similar asynchronous male reproductive pattern was reported for the tropical *Emballonuridae*, *Taphozous georgianus* [30], suggesting that other tropical species may present similar reproductive modifications to those seen in temperate zone species. On the other hand, in European members of the genus *Miniopterus*, whose male reproductive cycle does not overlap the dormancy period, activity of the secondary sex organs is completed before hibernation begins [22].

Gustafson [22] sustains that the asynchronous male reproductive pattern of bats can be sub-divided based on differences in the annual periodicity of the interstitial tissue. In all the species studied, the evolution of Leydig cells occurs during the period of spermatogenesis, producing differences related to the temporality of the involution of these interstitial cells [22]. Thus, in one sub-type observed in species like *Eptesicus serotinus*, *Nyctalus noctula*, *Pipistrellus herperus*, *P. pipistrellus*, *Rhinolophus ferrumequium* and *R. hipposoderus*, the Leydig cells persist in a developed state, or with only a slight reduction in activity during the periods of mating and hibernation; while complete involution occurs in the spring, followed by emergence from hibernation [22]. A second sub-type was identified in the species *Eptesicus fuscus*, *Myotis blythii*, *M. capaccini*, *M. grisescens*, *M. lucifugus*, *M. myotis*, *M. velifer* and *Corynorhinus townsendii*, where involution of the Leydig cells is completed at the onset of the mating period, or is likely kept at a low level of activity during the overlap between mating and hibernation [22]. In case of *Miniopterus schreibersii*, a species that does not present the asynchronous male reproductive pattern, involution of the Leydig cells is completed during the mating period, and these cells are inactive during hibernation [22].

Temporal Asynchrony between Male Reproductive Functions and Prolonged Sperm Storage

As mentioned above, some bat species, mainly of the *Vespertilionidae* and *Rhinolophidae* families, show significant temporal asynchrony in the development and functions of the male reproductive organs [15, 16, 23]. In these bats, testicular development and spermatogenesis transpire primarily in summer, while full development of the epididymis and accessory glands, the expression of libido, and mating, all occur in autumn. This asynchrony produces a long, highly uncommon, period of epididymal sperm storage that may last several months after full testicular regression. During this period, the sperm must be kept viable [12 - 14,

16, 31 - 33]. The process of prolonged sperm storage is defined as a condition in which the sperm remain viable *with* fertilizing capacity for long periods [34]. It is important to recall that, for mammals in general, including humans, sperm are stored in the epididymis for just 11 days [35]. Thus, the species with the longest sperm storage times include some bats. Studies have recorded instances of sperm being stored in the epididymis for various months, remaining viable for 4-10 months, depending on the species and the environmental conditions of their habitats [12 - 14, 16, 31 - 33].

Early works that sought to explain this important reproductive phenomenon argued that prolonged sperm storage could be a natural consequence of the hypothermia characteristic of hibernation. However, we know of many non-hibernating tropical species that present asynchrony and the long sperm storage in the epididymis this entails [36 - 38]. Also, some species of hibernating bats that store sperm wake regularly while hibernating [36]. A careful reading of those early studies shows that they focused exclusively on the processes involved in maintaining sperm viability [31, 32], paying scant attention to the changes related to ESM which, in most mammals, is independent of sperm storage because the sperms mature before reaching the caudal region of the epididymis [1, 4, 39]. However, in the *Corynorhinus mexicanus* bat of the *Vespertilionidae* family, final sperm maturation occurs *in* the caudal region and is dependent on the duration of sperm storage [33, 40, 41].

Finally, it is important to note that bats with prolonged sperm storage also show temporal asynchrony between the functions of the testicle (production of sperm and androgens) and the epididymis (maturation and sperm storage), because the latter organ is morphologically and functionally dependent on the androgens produced in the testicle. Therefore, the phenomenon of androgen-independent ESM regulation in these bats is of great interest.

EPIDIDYMAL SPERM MATURATION IN BATS

In vertebrates with internal fertilization, including bats, sperms must travel through the epididymal tubule to acquire the ability to fertilize [42]. As noted above, the process through which this is achieved is epididymal sperm maturation (ESM). This begins once the male gametes enter the epididymal tubule [43, 44].

Morphology and Histology of the Epididymis

Reports on the nature of the epididymis in bats show that, like all mammals, it consists of one sole, narrow, highly-convoluted tubule integrated into the male's excurrent tract that connects the efferent tubules to the deferent duct. The anatomy of the epididymis includes four main regions: a) the initial segment, which receives the sperm and fluid from the testicle and efferent vessels; b) the caput (head); c) the corpus (body), which contains two sub-regions where sperm maturation is reported to take place in most mammals; and, d) the cauda, where mature sperms are stored (Fig. **1**) [1, 40, 41, 45 - 53].

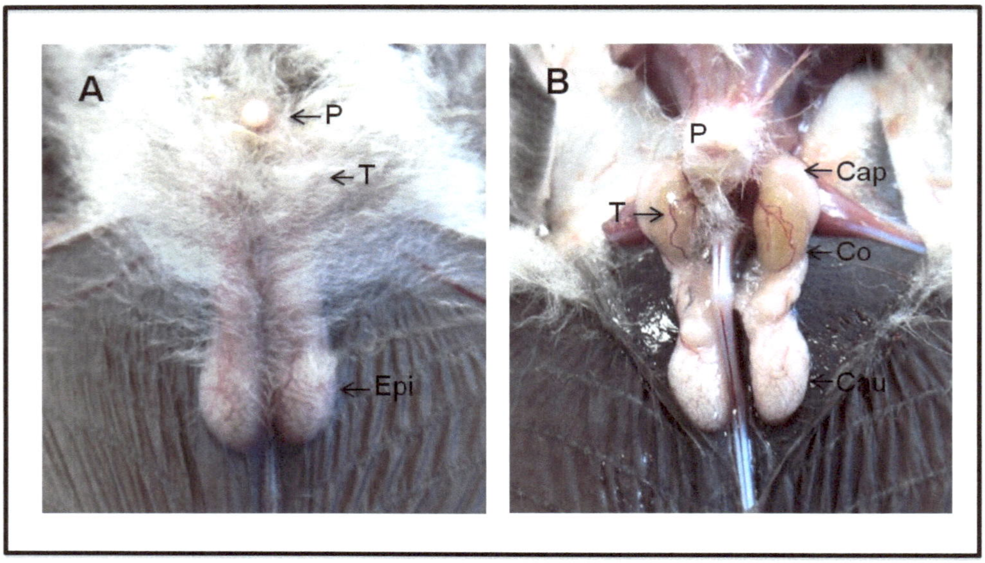

Fig. (1). Male sexual organs of the Mexican bat *Corynorhinus mexicanus* in mating season (December). A) external view of the genitals, it is appreciated the epididymal cauda well developed . B) internal appearance, it is noted the organization and regionalization. P) Penis; T) Testis; Ca) Caput; Co) Corpus; Cau) Cauda.

The epididymis of non-seasonally reproducing bat species is not significantly different in morphology from this organ in all other non-seasonal mammal species [51 - 53]. However, one significant variation in seasonal bats is this organ's clearly-marked periods of recrudescence and involution depending on the phase of the reproductive cycle. It appears distended to the naked eye when sperms are not present [54], but when sperm enter the caput, this region increases in size progressively as they move through the epididymal tubule, pass through the corpus, and settle in the cauda. In monoestrous species, sperms leave the anterior regions to be stored in the cauda. At that moment, this region's appearance is similar to that of a bulbous sac [55 - 57]. Reports on the *Corynorhinus mexicanus* bat indicate that involution of the caput occurs one month earlier than in the cauda [16, 57].

The different regions of the epididymal tubule play fundamental roles in controlling and inducing sperm maturation [6, 50, 58 - 60]. As the sperms traverse this organ, they come into direct contact with specific luminal microenvironments in each region, which are further sub-divided into intra-regional segments where the processes of gene expression and protein synthesis differ [1, 61, 62].

The histological characteristics of the epididymis have been reported for several mammal species, including bats, highlighting the peculiarities of the various segments [51, 57, 63, 64]. Each epididymal region is histologically-distinct from the others, and the presence of diverse cell types that effectuate the synthesis, secretion and absorption of the substances that come into contact with the sperm has been reported; namely, principle, basal, halo, clear, narrow, and apical cells (Fig. **2**) [1, 48, 64]. The principle cells are found throughout the epididymis, and in seasonal species are evident year-round; but this is not true for other cell types, as these may vary (also in distribution) depending on the region and reproductive season [51, 54, 57]. Studies have determined that the cells in the epididymal tubule form a hemato-epididymal barrier made up of tight bonds between adjacent principle cells of the epithelial tissue. This allows conservation of the luminal microenvironment for sperm maintenance by restricting the entry of numerous ions, solutes and macromolecules that cross the epididymal epithelium [1, 32, 45].

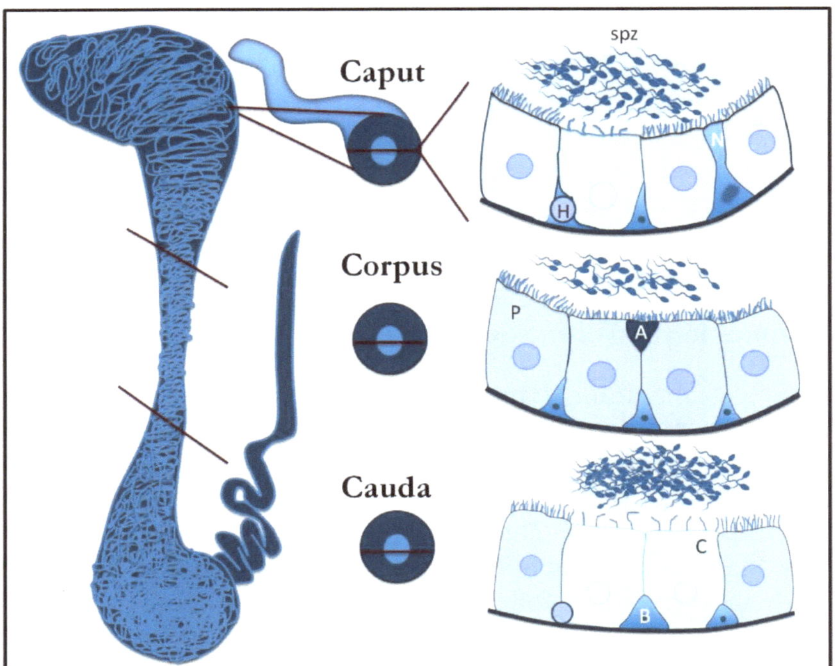

Fig. (2). Representative diagram of the three main anatomical regions of the epididymal tubule (Caput, Corpus and Cauda) of Mexican bat Corynorhinus mexicanus and histology. It is shown a cross section for each region of epididymis and the different cell types present in each region are outlined. A) Apical cell; B) Basal cell; C) Clear cell; H) Halo cell; N) Narrow cell; P) Principal cell.

One example of this was reported by Crichton [32] in species with prolonged sperm storage (*Myotis lucifugus*, *Myotis septentrionalis*) where the tight bonds in the epithelium of the caudal region of the epididymis are specialized to favor an increase in the solutes in the lumen of the tubule. This increase in osmolarity correlates with the period of sperm storage, since the movement of large quantities of solutes that are impermeable to the epithelium causes the sperm to dehydrate, reduce their metabolism and propitiate quiescence. This process may involve the AQP9 aquaporins, which participate significantly in transporting water and neutral solutes. This phenomenon has been studied in the seasonal *Artibeus lituratus* bat, where AQP9 was found throughout the epididymis, but primarily in the caudal region, and were expressed seasonally with a diminished presence as the epididymis regressed [65].

This description shows that the microenvironment generated by the epididymal epithelium favors differential functioning among the distinct regions. As a result, the ionic composition, organic molecules and proteins that contact the sperm as they move through the epididymis, produce the continuous modifications that permit: 1) an increase in the negative charge of the plasmatic membrane; 2) loss of the cytoplasmic droplet; 3) a reduction in the size of the head (associated with hyper-compaction of the chromatin); 4) a change in the phospholipid-cholesterol coefficient; 5) restructuring or remodeling of the shape of the acrosome; 6) a reduction in the diameter of the mitochondria; and, 7) increased progressive movement [43, 49, 66 - 74]. Together, these changes give the sperm the potential to become capacitated, undergo the acrosomal reaction and, finally, fertilize the oocyte [43, 49, 70, 72 - 74].

Incorporation of Proteins into Sperm and the Participation of Epididymisomes in the Maturation Process

Some of the most important elements of the microenvironment generated by the epithelium of the epididymis that favor the maturation process are the proteins synthetized and secreted in the epididymis, which control many processes related to maturation. Later, the incorporation of many other proteins will give the sperms their fertilization capability. These sperms are thus differentiated from those produced by the testicle, which are incapable of fertilization. Once synthetized and secreted, these proteins follow various pathways through the epididymis: some adhere to pre-existing acceptor sites on the sperm's surface, others incorporate antigens and use them to bond to the sperm, while others bond directly, although they are external to the sperm [75].

The incorporation of proteins from the epididymal epithelial tissue into the sperm takes place by region; a fact that has been demonstrated for species as diverse as bulls, pigs, rams, rodents and horses [59, 76]. There are marked changes between the sperm's tail, mid-section, and acrosomal regions (Fig. **3**) [77, 78].

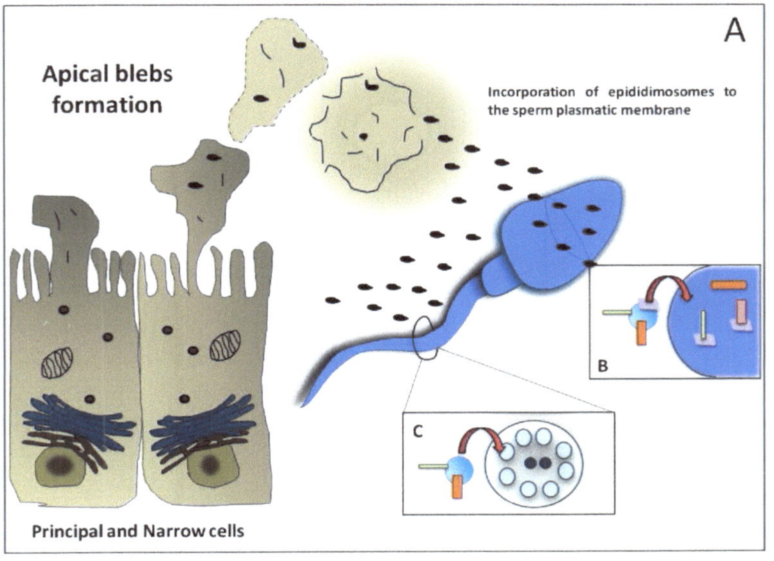

Fig. (3). Schematic representation of the formation of aposomas (epididimosomas) by the Principal and Narrow cells of the epididymal tubule in mammals (A). B) Proteins transported through of epididimosomas to the cephalic region of sperm. C) Transport of proteins through of epididimosomas inside of the axoneme of sperm.

It is well-established that protein secretion by the principle cells of the epididymal tubule occurs merocrinally [79], but some of the proteins that bond to the sperm as they pass through the epididymis do so by means of a glycosylhos-phatidylinositol-anchor protein, also called the 'major maturation antigen' [80 - 82], so alternate pathways for protein acquisition by the sperm have been posited [50, 83]. One such pathway involves apocrine secretions, whose presence has been reported in various species [84].

Reports indicate that one characteristic of the principle and narrow cells of many mammal species is the formation of bubbles in the apical section. Since these bubbles interact with the sperm as they pass through the epididymal tubule, it is believed that they also participate in the process of sperm maturation [79, 84].

Based on the classic process of exocrine secretion, diverse studies argue that the epithelial cells of the epididymal tubule may be involved in the processes of apocrine secretion [85], which propitiate the formation of a bubble in the apical section of the principle cells of the epididymal tissue. This bubble later detaches in the form of a vesicle that moves towards the lumen of the epididymal tubule [50, 83, 86]. These vesicles have been called aposomes, and are seen in the seminal plasma and lumen of the excurrent duct of the males of various mammal species, including rabbits, primates, sheep, dogs, cats, bovines, horses, rats and humans [5, 79, 87]. However, when found inside the epididymal tubule, they are called epididymisomes; *i.e.,* vesicles with diameters in the range of 10-100 μm [50, 83, 86, 88]. Epididymisomes originate in various regions of the epididymis and have differences that reflect their protein composition and the pathway through which they interact with, and transfer proteins to the sperm (Fig. **3**) [89 - 91].

Over the past ten years, researchers have sought to understand how aposomes interact with sperm, and the function(s) associated with them. There are reports that they possess fusogenic properties with the sperm membrane and so effectuate exchanges of material. *In vitro* studies have shown that aposomes do indeed transfer proteins from the epithelial tissue of the epididymal tubule to the sperm [5, 79, 87]. Thus, we would expect that the proteins acquired during maturation will occupy a strategic, functional location in the distinct structures of the sperm. Later, when many of those proteins are transferred to the sperm, they may either bond to the sperm membrane [80] or be incorporated into the internal structures of the cell [89]. The proteins secreted by the caput and corpus regions of the epididymis are involved in the acquisition of notable motility and the ability to bond to the pellucid zone and so fuse with the plasmatic membrane of the oocyte, as well as in modulating sperm motility in the epididymis and immunological fertilization. Moreover, they protect against oxidative stress and eliminate defective sperm [49, 61, 79, 87]. Finally, the proteins secreted in the cauda are required to maintain the viability and fertilization potential of the stored sperm [79].

As mentioned above, the *Corynorhinus mexicanus* bat has prolonged sperm storage associated, in this case, with ESM in the cauda of the epididymis. The formation of vesicles in principle cells has been verified, together with evidence of secretor granules (Fig. **4**) that may be epididymisomes. Also as noted previously [4], vesicle formation is related to ESM [57]. These changes are reflected in the Golgi apparatus, the secretor granules, and the endocytic apparatus (*i.e.*, vesicles, endosomes, multi-vesicular bodies and lysosomes) (Fig. **4**) [56].

However, while many proteins are synthetized in, and secreted by, the epididymis before incorporation into the sperm, the largest amount of proteins that form part of the fertile sperm is synthetized in the testicle. Therefore, modification of the sperm proteins on the path towards maturation involves other important processes.

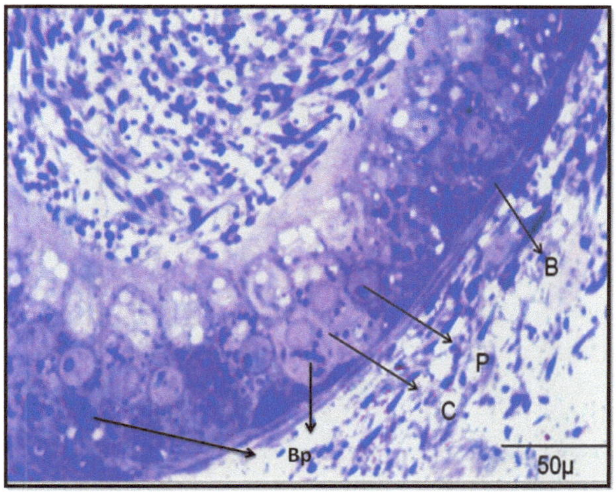

Fig. (4). Photomicrograph of cross section of the caudal region of the epididymis of Corynorhinus mexicanus; September. B) Basal cell; C) Clear cell; P) Principal cell; Bp) Blebs production. Staining: Toluidine blue.

Modification of Sperm Proteins During Maturation

Protein Glycosylation

Sperms are coated by a plasmatic membrane, characterized mainly by its composition of lipids and glycoproteins, though it also contains a small amount of cytoplasm [47, 78]. It is suggested that the external membrane of the sperm is highly-glycosylated to form what is called the glycocalyx, a name derived from the large quantities of proteins that are clearly glycosylated. But not all molecules are strictly bonded to the membrane surface, as some are intercalated with phospholipids on the membrane, while others are found superficially in association with the membrane through polar groups or hydrophobic interactions [81].

Numerous studies have determined that the surface of the sperm membrane is not homogeneous, but contains distinct functional domains –the acrosome, the equatorial segment, the post-acrosome, the mid-piece, and the principle piece– that produce a characteristic spatial distribution of the elements in the glycocalyx [80, 92]. This highly-heterogeneous micro-geography of the sperm membrane holds part of the key for understanding the complex functions of sperm during their post-testicular existence [47, 78, 81, 93 - 95].

Now that the role of the carbohydrate portion of the proteins is better understood, glycosylation is considered one of the post-translational modifications present in the surface proteins of the sperm. This implies multiple intra- and extracellular biochemical modifications related to maturation and storage in the epididymis up to ejaculation, as well as the adhesion of the sperm to the pellucid zone of the oocyte [75, 92, 96, 97].

When the sperms leave the testicle, they are also covered by various antigens. As they cross the epididymis, these antigens may be absorbed, altered or lost [93]. As noted previously, diverse cell types inside the epididymal lumen secrete proteins that act intensely to modify the sperm [98]. As is the case of other cell types, the maintenance, maturation and permanence of sperm inside the epididymis require energy, so the participation of glycoproteins from the epithelial tissue that forms the epididymal duct is indispensable. Later, these glycoproteins come into contact with the sperm [99, 100].

Reports indicate the presence of two groups of enzymes that function to modify the carbohydrates on the sperm's surface: 1) glycohydrolases: hydrolytic enzymes that break down sugar residues in the existing glycoconjugates; and, 2) glycosyltransferases: synthetic enzymes that add sugar residues from other sugar donors (nucleotides) to an acceptor site on the existing molecules, which are present in high concentrations in the luminal fluid of the epididymis that surrounds the sperm. Both groups play roles in modifying the glycoproteins on the sperm's surface during ESM [96]. As this process develops, molecules that contain such carbohydrate chains as D-galactosidase and N-acetyl-D-galactosamine reduce their molecular size during glycosylation and/or proteolysis due to the replacement of several of their glycoconjugates [92].

Studies using specific glucosidases, such as exoglycosidases and glycoprotein synthesis inhibitors, suggest that a wide variety of oligosaccharides are synthetized by bonding to the lipids (N-ligand) present on the sperm's surface, and are involved in fertilization in mammals [101]. Another characteristic associated with sperm maturation also found in most mammals is the general increase in the negative charge on the surface of the plasmatic membrane [93, 102] caused by incorporating sialic acid from the epithelial cells of the epididymis, or from antigens that increase the acidity of the medium where they exist [103 - 106].

Carnitine and glycerophosphocholine are two glycoproteins incorporated into the epididymis. Supplied by the epididymal fluid [99], they function as energy sources for sperm metabolism [100]. Alteration of the characteristics of the pre-existing enzymes allows the addition of new proteins from the plasmatic membrane of the sperm mediated –at least partially– by enzymes like galactosyltransferase and sialyltransferase, which are also present in the epididymal fluid, as well as the α-lactalbumin that regulates protein glycosylation on the sperm's surface [78].

These findings allow us to evaluate the membrane carbohydrates of the sperm and the use of lectins that bond to specific carbohydrates present in the membrane, and to determine their quantity and distribution [107, 108]. Thus, we can now determine the presence of N-acetyl glucosamine and/or sialic acid, fucose and mannose in the epididymal sperm of the *Corynorhinus mexicanus* bat, and verify that the N-acetylglucosamine carbohydrates and/or sialic acid change their distribution patterns: *i.e.*, instead of being found in the post-acrosomal region of the sperm, they are distributed throughout the sperms in the mature state. The opposite occurs with mannose distribution: instead of being distributed throughout the sperm, it is found only in the sub-acrosomal region of mature sperm. However, it is important to note that the amounts of sugars, N-acetylglucosamine and/or sialic acid, fucose and mannose decrease as the sperm approach maturity [109].

Protein Phosphorylation

During ESM, in addition to the changes mentioned above, the amount of cyclic adenosine monophosphate (AMPc) increases in association with a change in the distribution and quantity of phosphorylation of tyrosine residues in the proteins. These changes are essential for the coordinated development of movement (among other functions [107 - 110]) since the maturation process includes a step called protein-activation that is characterized by tyrosine phosphorylation. In some cases, studies have determined that this is due to a kinase-dependent tyrosine of AMPc [9, 71, 111]. In addition to the activity of kinase proteins, the state of phosphorylation is controlled by the activity of phosphatases: the former aggregate to the phosphate group, while the latter function to remove it [9, 71, 111].

It has been shown that the events of phosphorylation/dephosphorylation allow the sperms to turn their function on or off, like a "switch", mainly in the proteins of the sperm's flagella, by participating in the acquisition of motility and the capacity to fertilize the oocyte [68, 71].

In order to understand how the cascade of signalings of tyrosine phosphorylation is activated, we must keep in mind that the protein adenylate cyclase is the first mediator in performing protein phosphorylation, a process that increases AMPc levels. Also, this is the second messenger in all cell types that activates the kinase A protein (PKA) which, in turn, activates the tyrosine kinase protein [9, 68, 71, 112].

The process of tyrosine phosphorylation in the sperm proteins associated with ESM has also been studied in the *Corynorhinus mexicanus* bat. There, it has been determined that tyrosine phosphorylation of its proteins increases markedly prior to the period in which sperm acquire the potential to become capacitated and perform the acrosomal reaction [37]; two clear indicators of mature sperm. But secretion of glycoproteins, glycolipids, phospholipids –all the characteristic of ESM– is enabled by gene expression in a certain zone of the epididymis. This expression is regulated mainly by androgens [113], and is a topic of great interest given the reproductive peculiarities of seasonal bats that show asynchrony between the functions of the testicle and epididymis.

REGULATION OF THE MATURATION PROCESS

Regulation by Androgens

Studies have determined that after surgical castration in rats, the epididymis tends to decrease in weight, luminal diameter, and the height of the epithelium in the epididymal tubule. Also, the secretion and absorption of epithelial cells are compromised, because these changes are accompanied by the loss of microvilli on the apical surface [110]. Additionally, observations indicate that the principle cells are the most sensitive of all cell types in the absence of androgens due to a drastic reduction of the smooth endoplasmic reticulum and the Golgi apparatus, as well as an accumulation of lysosomes, a vacuolization process, and the disappearance of vesicles [110]. Other effects of castration include the induction of cell death in the epididymis by apoptosis, beginning in the initial segment, followed by the other regions up to the cauda. Also, sperms fail to acquire motility, lose the capacity to fertilize the oocyte, and die [111, 112]. However, when testosterone was administered to the castrated rats, most of the histological characteristics of the epithelium were restored, except in the initial segment, which is highly-dependent on non-hormonal factors from the testicle [110]. Therefore, it is argued that the normal function of the epididymis is governed by androgens and testicular factors that regulate the functioning of the initial segment and the mechanisms through which this operates.

Testosterone synthesis is regulated by the luteinizing hormone (LH) secreted by the adenohypophysis. Upon reaching the Leydig cells (interstitial cells in the testicle), LH regulates testosterone synthesis. On the other hand, the follicle-stimulating hormone (FSH), also secreted by the adenohypophysis, regulates secretion of the androgen-binding protein (ABP), which is controlled by the Sertoli cells (inside the seminiferous tubule). Once bonded to ABP, testosterone is transported through the lumen of the seminiferous, *rete testis* and efferent tubules until it reaches the epididymal tubule, where it is bio-transformed through the action of the enzyme 5-alpha-reductase into dihydrotestosterone, an androgen more powerful than testosterone (Fig. **5**).

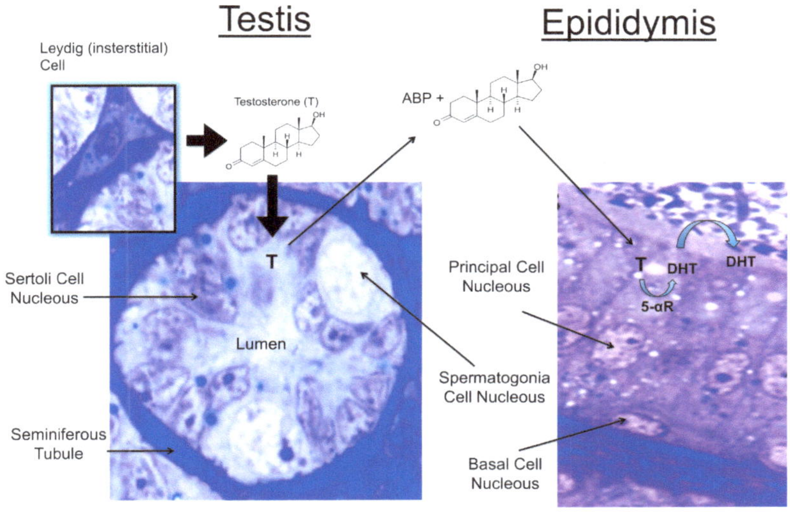

Fig. (5). Scheme of the participation of the testosterone on the functionality of the epididymis. Testosterone (T) produced in Leydig cells, is transported inside the seminiferous tubules. Androgen binding protein (ABP) carries testosterone to the Principal cells of the epididymis to maintain full operation of the same.

The epididymis' dependence on the testicle is due to the fact that testosterone is transported through the lumen of the tubules. This is the origin of the lumicrine term [113], which refers to the fact that testicular factors are required to maintain the structure and function of the epididymis, and that control of the epididymal function may be exercised, in part, by androgens of testicular origin, together with non-hormonal factors present in the liquid from the *rete testis*. The action of the androgens on the epididymis occurs primarily through gene expression, largely in the corpus and cauda regions [111, 114], where the genes involved include cystatin, 5 alpha reductase, and gamma glutamyl transpeptidase.

In non-seasonal polyestrous bats, regulation of the epididymis by the testicle occurs without any changes when compared to the elements reported for other non-seasonal mammals, since testicular activity (synthesis of androgens and sperm), is continuous. But, what happens in the particular case of bats that experience testicular regression while maintaining the presence of sperm?

This has been determined in bats of the *Vespertilionidae* family. For example, the species *Myotis ricketti* presents a stage of testicular involution that coincides with the period of sperm storage. Observations show that testosterone concentrations decrease dramatically precisely in this stage, indicating that testosterone does not maintain sperm viability [54]. Males of the species *Myotis nigricans* present two periods of total testicular regression during the same annual reproductive cycle. Sperm storage occurs during the first stage of testicular regression, but not the second. The patterns of hormonal regulation in the regression period associated with the presence of sperm show that despite a clear reduction in testicular testosterone production, androgen receptors are present in the corpus and cauda regions (presence of sperm), but not the caput (absence of sperm), suggesting that in spite of the reduced testosterone concentration available, the epididymis maintains its functionality and propitiates sperm maturation [49]. Those authors suggest that the functionality of the epididymis is maintained because large quantities of testosterone accumulate in the lumen of the epididymal tubule, but this issue has not been fully clarified, so it is important to explore other potential regulating mechanisms as we seek to explain the mechanisms involved in maturation.

Regulation by Reactive Oxygen Species

One little-explored, but well-established, regulating mechanism in the ESM process is oxidation of biomolecules by reactive oxygen species (ROS). ROS are produced by metabolism through the participation of oxygen (O_2). Generally-speaking, they result from the excitation of O_2 to form singlet oxygen, or through the transfer of 1, 2 or 3 electrons to O_2 to form the superoxide anion ($^{\cdot}O2^{-}$), hydrogen peroxide (H_2O_2), and the radical hydroxyl ($^{\cdot}OH^{-}$), respectively [115, 116]. These ROS, except for H_2O_2, are free radicals; that is, molecules with unpaired electrons that avidly seek to complete the electron pair so as to neutralize electron spin (s = +1/2 ó -1/2) and permit oxidization in the form of lipids, carbohydrates, proteins and nucleotides through structural modification. These alterations can cause cell damage –as occurs in lipid peroxidation (oxidation of fatty acids in the membrane)– or activate/inhibit proteins in a well-controlled process that may foster physiological processes, as occurs with sperm (Fig. **6**).

A key piece in controlling ROS production is the primary defense system, which is made up of the anti-oxidizing enzymes superoxide dismutase (SOD), glutathione peroxidase (GPX), and catalase (CAT). The enzyme SOD catalyzes the $.O_2^-$ in H_2O_2 and water; while GPX is a selenoprotein that catalyzes the decomposition of H_2O_2 using reduced glutathione as the substrate to produce oxidized glutathione plus water. Finally, CAT also catalyzes decomposition of the H_2O_2 in O_2 and water [117]. While H_2O_2 is not a radical, it is the precursor of one of the most dangerous of all radicals; the one responsible for lipid peroxidation: OH^-.

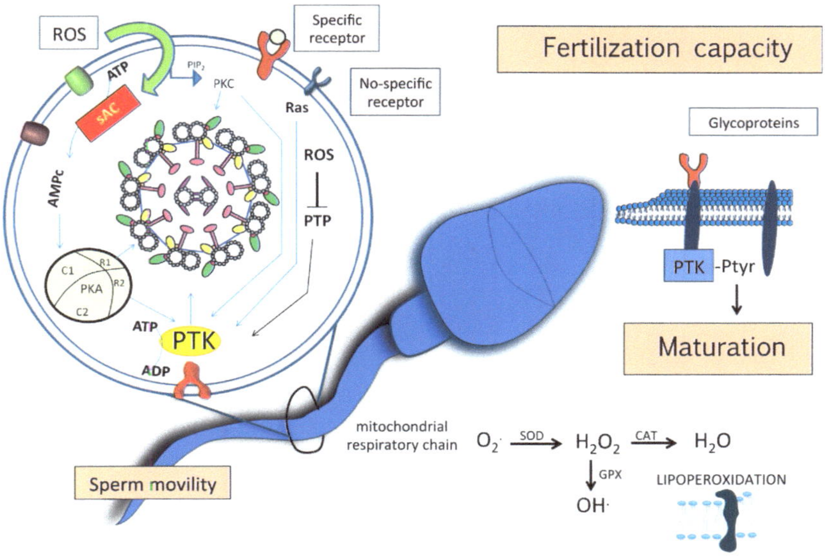

Fig. (6). Reactive oxygen species participation in sperm physiology. Glycoproteins found in the sperm plasma membrane are essential for activation of different signaling pathways that trigger of tyrosine phosphorylation, promoting the maturation of sperm, and with this, their potential for fertilization.

Research also shows that ROS are a key factor in the process of sperm maturation, and it is now clear that they are fundamental for such physiological processes in the sperm as maturation and capacitation [70, 116, 118 - 122]. It has been suggested that ROS are involved in initiating sperm motility by increasing AMPc synthesis and protein phosphorylation [71, 117, 118, 123 - 125]. In some cell types, the increase in tyrosine phosphorylation results from the oxidation and direct activation of the kinase proteins [126, 127]. However, observations show that the active center of the phosphatases contains a large amount of cysteine residues, which are essential for their activity [128, 129]. Hence, the inhibitory effects of the anti-oxidants in the activity of the phosphatases have been considered the most probable mechanism directly responsible for tyrosine phosphorylation [129 - 131].

Studies on the *Corynorhinus mexicanus* bat have determined that ROS production in epididymal sperm may play a prominent role in the ESM, since increases and reductions in the amounts of ROS were found to be perfectly-regulated with the increase and reduction of the activity of the anti-oxidizing enzymes SOD, GPX and CAT [41]. This suggests the existence of a change in the equilibrium redox that, in this species, is closely-associated with sperm maturation and storage.

Clearly, the process of epididymal sperm maturation in bats has been inadequately studied, as most published research focused on reproductive patterns and testicular physiology, while largely ignoring the epididymis. The vast majority of published works on the physiology of the epididymis associated with ESM is by our work group and is based on the *Corynorhinus mexicanus* bat of the *Vespertilionidae* family. Thus, we argue that it is important to integrate information on the ESM process in this bat species.

THE *CORYNORHINUS MEXICANUS* BAT AS A MODEL FOR STUDYING THE ESM PROCESS

The *Corynorhinus mexicanus* bat is an endemic species in Mexico that inhabits the highest, most humid zones of the main mountain systems, habitually-localized above 2000 meters above the sea level. The reproductive pattern of the female of this species is monoestrous, but males have only one reproductive cycle per year, marked by temporal asynchrony between the primary and secondary reproductive functions. The testicles reach maximum size in August, while maximum development of the epididymis and the group of accessory sex glands occurs 1-3 months later. This means that the sperm must be stored in the epididymis for as long as 4 months [15, 16, 55, 132], as was confirmed by examining sperms obtained from the epididymis of these bats throughout their reproductive cycle. Sperm cells are present from August to January, but a clear reduction in concentrations is evident in the cephalic region in late October, while a clear increase occurs in the caudal region [33, 55, 133].

Sperms taken from the cauda in September showed large amounts of cytoplasmic droplet and a low potential to become capacitated and achieve the acrosomal reactions, two clear indicators of immaturity. But these change as the period of sperm storage advances, so that by late October, the sperm acquire the potential to become capacitated and perform the acrosomal reaction, accompanied by a reduced occurrence of cytoplasmic droplet (Fig. **7**) [33].

This process appears to be regulated by post-translational modifications of the proteins that include phosphorylation, glycosylation and, possibly, oxidation. Studies show that this species has an increase in tyrosine phosphorylation of the proteins and in the concentration of ROS in sperms obtained in September and October, respectively. However, in contrast to reports on other species and processes [40, 41], the participation of ROS does not seem to be related to the activation of cyclase adenylate, since protein phosphorylation (where it is related to cyclase adenylate activity), does not coincide with the most significant increase in the amount of ROS [40, 41]. This leads us to think that the ROS may participate in a pathway distinct from tyrosine phosphorylation; perhaps by activating/inhibiting proteins through oxidization. But this has not been confirmed.

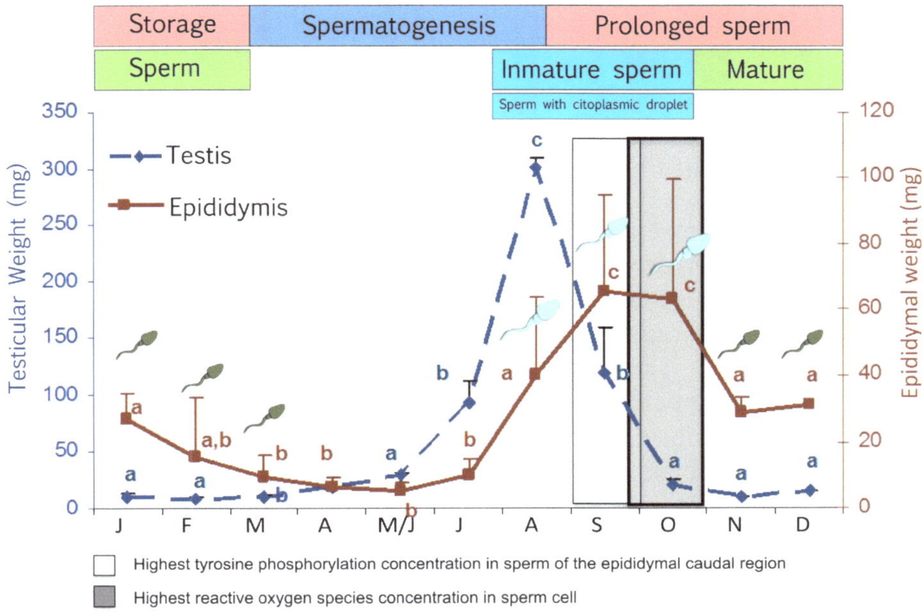

Fig. (7). Graphic representation of the male reproductive activity of the Corynorhinus mexicanus bat. The weight of the testicle (blue dotted line) and the epididymis (continuous red line) of bats captured monthly, for one year. The upper bar in gray represents the period of spermatogenesis and white upper bar, represents the storage sperm period. The turquoise sperm cells, represent immature spermatozoa, the olive green sperm, represent the mature spermatozoa. The regulators of sperm maturation are represented as follows: in vertical rectangle white color: maximum values of protein phosphorylation in tyrosine; In vertical gray rectangle: maximum concentration values of reactive oxygen species; Blue horizontal rectangle: maximum values of N-acetyl glucosamine and sialic acid; Horizontal rectangle red color: minimum values of N-acetyl glucosamine and sialic acid.

What is clear is that on the basis of the events described herein, we will find an increase in the amount of mature sperm, meaning that these processes are indispensable in the process of sperm maturation in *Corynorhinus mexicanus* bats. Also, modifications of the carbohydrates in the sperm membrane (N-acetylglucosamine and/or sialic acid, fucose and mannose) seem to be important in the maturation process [109], though their precise role is still unclear.

CONCLUSION AND PERSPECTIVES

The studies conducted to analyze the process of epididymal sperm maturation in *Corynorhinus mexicanus* bats provide a new perspective on a phenomenon that has been widely-studied in non-chiroptera mammals, where it seems clear that in most species sperms mature before reaching the caudal region. But this is not the case in this bat species where, in addition to finding that sperm maturation ends in the region identified heretofore as "exclusively" for sperm storage, maturation is dependent on the period during which the sperms are stored in this region. Now, the question is: can these studies be applied to all species that present prolonged sperm storage?

Finally, the following aspects of the maturation process in species with asynchrony between testicular and epididymal functions need to be clarified:

- The question of the absence of androgens/presence of receptors, and their role in epididymal functionality; including: protein synthesis linked to sperm maturation and storage not associated with testicular functioning.
- The role of proteins transported by epididymisomes: what are they and how do they act?
- Alternative signaling pathways to adenylate cyclase, and the factors that activate it for tyrosine phosphorylation.
- The role of ROS in relation to the maturation process.
- The status of the membrane lipids linked to the maturation process.

CONFLICT OF INTEREST

The authors declare no conflict of interest, financial or otherwise.

ACKNOWLEDGEMENTS

We are thankful for the help given to our research group from the Morphophysiology and Biochemistry of Sperm (UAM-I) and especially Mario Gabriel Hidalgo.

REFERENCES

[1] Robaire B, Hinton BT, Orgebin-Crist MC. The Epididymis. In: Neill JD, Wassarman P, Eds. Knobil and Neill's Physiology of Reproduction 1. 3rd ed. USA: Academic Press 2006; pp. 1071-148. [http://dx.doi.org/10.1016/B978-012515400-0/50027-0]

[2] Robaire B, Hamzeh M. Androgen action in the epididymis. J Androl 2011; 32(6): 592-9. [http://dx.doi.org/10.2164/jandrol.111.014266] [PMID: 21764895]

[3] Hermo L. Structural features and functions of principal cells of the intermediate zone of the epididymis of adult rats. Anat Rec 1995; 242(4): 515-30. [http://dx.doi.org/10.1002/ar.1092420408] [PMID: 7486022]

[4] Hermo L, Robaire B. Epididymal cell types and their functions. In: Robaire B, Hinton BT, Eds. The Epididymis: From Molecules to Clinical Practice: A Comprehensive Survey of the Efferent Ducts, the Epididymis and the Vas Deferens Boston, MA. Springer, US 2002; pp. 81-102.

[5] Girouard J, Frenette G, Sullivan R. Compartmentalization of proteins in epididymosomes coordinates the association of epididymal proteins with the different functional structures of bovine spermatozoa. Biol Reprod 2009; 80(5): 965-72. [http://dx.doi.org/10.1095/biolreprod.108.073551] [PMID: 19164173]

[6] Dacheux JL, Belleannée C, Guyonnet B, *et al.* The contribution of proteomics to understanding epididymal maturation of mammalian spermatozoa. Syst Biol Reprod Med 2012; 58(4): 197-210. [http://dx.doi.org/10.3109/19396368.2012.663233] [PMID: 22788532]

[7] De Vries M, Ramos L, Housein Z, De Boer P. Chromatin remodelling initiation during human spermiogenesis. Biol Open 2012; 1(5): 446-57. [http://dx.doi.org/10.1242/bio.2012844] [PMID: 23213436]

[8] Caballero J, Frenette G, Sullivan R. Post testicular sperm maturational changes in the bull: important role of the epididymosomes and prostasomes. Veterinary medicine international 2010; 2011: 757194.

[9] Tardif S, Dubé C, Chevalier S, Bailey JL. Capacitation is associated with tyrosine phosphorylation and tyrosine kinase-like activity of pig sperm proteins. Biol Reprod 2001; 65(3): 784-92. [http://dx.doi.org/10.1095/biolreprod65.3.784] [PMID: 11514342]

[10] Bedford JM. Significance of the need for sperm capacitation before fertilization in eutherian mammals. Biol Reprod 1983; 28(1): 108-20. [http://dx.doi.org/10.1095/biolreprod28.1.108] [PMID: 6338941]

[11] Visconti PE, Bailey JL, Moore GD, Pan D, Olds-Clarke P, Kopf GS. Capacitation of mouse spermatozoa. I. Correlation between the capacitation state and protein tyrosine phosphorylation. Development 1995; 121(4): 1129-37. [PMID: 7743926]

[12] Racey PA, Entwistle AC. Life-history and reproductive strategies of bats. In: Crichton EG, Krutzsch PH, Eds. Reproductive Biology of Bats. Academic Press 2000; pp. 364-414. [http://dx.doi.org/10.1016/B978-012195670-7/50010-2]

[13] Crichton EG. Sperm storage and fertilization. In: Crichton EG, Krutzsch PH, Eds. Reproductive Biology of Bats. Academic Press 2000; pp. 295-320. [http://dx.doi.org/10.1016/B978-012195670-7/50008-4]

[14] Racey PA. The prolonged storage and survival of spermatozoa in Chiroptera. J Reprod Fertil 1979; 56(1): 391-402. [http://dx.doi.org/10.1530/jrf.0.0560391] [PMID: 381655]

[15] Arenas-Ríos E, León-Galván MA, Mercado PE, Rosado A. Superoxide dismutase, catalase, and glutathione peroxidase during epididymal maturation and prolonged storage of spermatozoa in the Mexican big-eared bat (Corynorhinus mexicanus). Can J Zool 2005; 83(12): 1556-65. [http://dx.doi.org/10.1139/z05-152]

[16] León-Galván MA, López-Wilchis R, Hernández-Pérez O, Arenas-Ríos E, Rosado A. Male reproductive cycle of Mexican big-eared bats, Corynorhinus mexicanus (Chiroptera: Vespertilionidae). Southwest Nat 2005; 50(4): 453-60. [http://dx.doi.org/10.1894/0038-4909(2005)050[0453:MRCOMB]2.0.CO;2]

[17] Arenas-Ríos E, León-Galván MA, López-Wilchis R, Rosado A. Regulators in epididymal maturation and prolonged storage sperm processes in the bat Corynorhinus mexicanus. In: Zupan JL, Mlakar SL, Eds. Bats: Biology, Behavior, and Conservation. Nova Science Publishers 2011; pp. 275-97.

[18] Wimsatt W. Some problems of reproduction in relation to hibernation in bats. Bulletin of the Museum of Comparative Zoology Harvard 1960; 124: 249-67.

[19] Ruf T, Geiser F. Daily torpor and hibernation in birds and mammals. Biol Rev Camb Philos Soc 2015; 90(3): 891-926. [http://dx.doi.org/10.1111/brv.12137] [PMID: 25123049]

[20] Sinha Hikim AP, Bartke A, Russell LD. Morphometric studies on hamster testes in gonadally active and inactive states: light microscope findings. Biol Reprod 1988; 39(5): 1225-37. [http://dx.doi.org/10.1095/biolreprod39.5.1225] [PMID: 3219392]

[21] Clarke JR. Physiological problems of seasonal breeding in eutherian mammals. In: Finn CA, Ed.

Oxford Reviews of Reproductive Biology 3. USA: Oxford University Press 1981; pp. 244-312.

[22] Gustafson AW. Male reproductive patterns in hibernating bats. J Reprod Fertil 1979; 56(1): 317-31. [http://dx.doi.org/10.1530/jrf.0.0560317] [PMID: 381650]

[23] Krutzsch PH. Anatomy, physiology and cyclicity of the male reproductive tract. In: Crichton EG, Krutzsch PH, Eds. Reproductive Biology of Bats. Academic Press 2000; pp. 91-155. [http://dx.doi.org/10.1016/B978-012195670-7/50005-9]

[24] Oxberry BA. Female reproductive patterns in hibernating bats. J Reprod Fertil 1979; 56(1): 359-67. [http://dx.doi.org/10.1530/jrf.0.0560359] [PMID: 381653]

[25] Krutzsch PH. Male reproductive patterns in nonhibernating bats. J Reprod Fertil 1979; 56(1): 333-44. [http://dx.doi.org/10.1530/jrf.0.0560333] [PMID: 381651]

[26] Racey PA. Ecology of Bat Reproduction. In: Kunz TH, Ed. Ecology of Bats Boston, MA. Springer, US 1982; pp. 57-104. [http://dx.doi.org/10.1007/978-1-4613-3421-7_2]

[27] Wimsatt WA. Some interrelations of reproduction and hibernation in mammals. Symp Soc Exp Biol 1969; 23: 511-49. [PMID: 5367180]

[28] Thomas DW, Fenton MB, Robert MR. Social Behavior of the Little Brown Bat, Myotis lucifugus: I. Mating Behavior. Behav Ecol Sociobiol 1979; 6(2): 129-36. [http://dx.doi.org/10.1007/BF00292559]

[29] Crews D. Gamete production, sex hormone secretion, and mating behavior uncoupled. Horm Behav 1984; 18(1): 22-8. [http://dx.doi.org/10.1016/0018-506X(84)90047-3] [PMID: 6538547]

[30] Jolly SE, Blackshaw AW. Prolonged epididymal sperm storage, and the temporal dissociation of testicular and accessory gland activity in the common sheath-tail bat, Taphozous georgianus, of tropical Australia. J Reprod Fertil 1987; 81(1): 205-11. [http://dx.doi.org/10.1530/jrf.0.0810205] [PMID: 3668950]

[31] Crichton EG, Krutzsch PH, Yanagimachi R. Stability of the sperm plasma membrane of hibernating bats (Myotis velifer) compared with other mammals. J Reprod Fertil 1993; 97(1): 1-4. [http://dx.doi.org/10.1530/jrf.0.0970001] [PMID: 8463999]

[32] Crichton EG, Hinton BT, Pallone TL, Hammerstedt RH. Hyperosmolality and sperm storage in hibernating bats: prolongation of sperm life by dehydration. The American journal of physiology 1994; 267(5-2): R1363-70.

[33] Cervantes MI, Arenas-Rios E, León-Galván MA, López-Wilchis R, Ambriz D, Rosado A. Spermatozoa epididymal maturation in the Mexican big-eared bat (Corynorhinus mexicanus). Syst Biol Reprod Med 2008; 54(4-5): 196-204. [http://dx.doi.org/10.1080/19396360802334466] [PMID: 18942027]

[34] Racey PA. Viability of bat spermatozoa after prolonged storage in the epididymis. J Reprod Fertil 1972; 28(2): 309-11. [http://dx.doi.org/10.1530/jrf.0.0280309] [PMID: 5058271]

[35] Rowley MJ, Teshima F, Heller CG. Duration of transit of spermatozoa through the human male ductular system. Fertil Steril 1970; 21(5): 390-6. [http://dx.doi.org/10.1016/S0015-0282(16)37502-1] [PMID: 5508505]

[36] Gopalakrishna A, Bhatia D. Storage of spermatozoa in the epididymis of the bat Hipposideros speoris (Schneider). Curr Sci 1980; 49(24): 951-3.

[37] Karim KB, Banerjee S. Storage of spermatozoa in the epididymis of the tropical bat, Rhinopoma hardwickei hardwickei (Gray). Anat Rec 1985; 211(3): 94A-6A.

[38] Singh K, Krishna A. Inhibitory effects of melatonin on testosterone but not on androstenedione

production during winter in the vespertilionid bat, Scotophilus heathi. J Pineal Res 1995; 19(3): 127-32.
[http://dx.doi.org/10.1111/j.1600-079X.1995.tb00181.x] [PMID: 8750346]

[39] Robaire B, Viger RS. Regulation of epididymal epithelial cell functions. Biol Reprod 1995; 52(2): 226-36.
[http://dx.doi.org/10.1095/biolreprod52.2.226] [PMID: 7711192]

[40] Rodríguez-Tobón A, Fierro R, León-Galván MA, Rosado A, Cortés-Barberena E, Arenas-Ríos E. Tyrosine phosphorylation as evidence of epididymal cauda participation in the sperm maturation process of Corynorhinus mexicanus bat. Acta Zoologica (stockholm) 2015; 1-9.

[41] Arenas-Ríos E, Rosado García A, Cortés-Barberena E, *et al.* Reactive oxygen species production and antioxidant enzyme activity during epididymal sperm maturation in Corynorhinus mexicanus bats. Reprod Biol 2016; 16(1): 78-86.
[http://dx.doi.org/10.1016/j.repbio.2016.01.001] [PMID: 26952757]

[42] Cooper TG. Sperm maturation in the epididymis: a new look at an old problem. Asian J Androl 2007; 9(4): 533-9.
[http://dx.doi.org/10.1111/j.1745-7262.2007.00285.x] [PMID: 17589792]

[43] Fàbrega A, Puigmulé M, Yeste M, Casas I, Bonet S, Pinart E. Impact of epididymal maturation, ejaculation and *in vitro* capacitation on tyrosine phosphorylation patterns exhibited of boar (Sus domesticus) spermatozoa. Theriogenology 2011; 76(7): 1356-66.
[http://dx.doi.org/10.1016/j.theriogenology.2011.06.007] [PMID: 21835448]

[44] Fàbrega A, Puigmulé M, Bonet S, Pinart E. Epididymal maturation and ejaculation are key events for further *in vitro* capacitation of boar spermatozoa. Theriogenology 2012; 78(4): 867-77.
[http://dx.doi.org/10.1016/j.theriogenology.2012.03.039] [PMID: 22704392]

[45] Serre V, Robaire B. Segment-specific morphological changes in aging Brown Norway rat epididymis. Biol Reprod 1998; 58(2): 497-513.
[http://dx.doi.org/10.1095/biolreprod58.2.497] [PMID: 9475407]

[46] Cooper TG. Interactions between epididymal secretions and spermatozoa. J Reprod Fertil Suppl 1998; 53: 119-36.
[PMID: 10645272]

[47] Eddy EM. Neill JD, Wassarman P. The Spermatozoon. Knobil and Neill's Physiology of Reproduction. 3rd ed.. USA: Academic Press 2006; pp. 3-54.
[http://dx.doi.org/10.1016/B978-012515400-0/50006-3]

[48] Setchell BP, Breed WG. Neill JD, Wassarman P. Anatomy, vasculature, and innervation of the male reproductive tract. Knobil and Neill's Physiology of Reproduction. 3rd ed.. USA: Academic Press 2006; pp. 587-96.

[49] Sullivan R, Frenette G, Girouard J. Epididymosomes are involved in the acquisition of new sperm proteins during epididymal transit. Asian J Androl 2007; 9(4): 483-91.
[http://dx.doi.org/10.1111/j.1745-7262.2007.00281.x] [PMID: 17589785]

[50] Sullivan R, D'Amours O, Caballero J, Belleannee C, Eds. The sperm journey in the excurrent duct: functions of microvesicles on sperm maturation and gene expression along the epididymis. Proceedings of the V International Symposium on Animal Biology of Reproduction.

[51] Beguelini MR, Góes RM, Rahal P, Morielle-Versute E, Taboga SR. Impact of the processes of total testicular regression and recrudescence on the epididymal physiology of the bat myotis nigricans (Chiroptera: Vespertilionidae). PLoS One 2015; 10(6): e0128484.
[http://dx.doi.org/10.1371/journal.pone.0128484] [PMID: 26057377]

[52] Beguelini MR, Puga CC, Martins FF, Betoli AH, Taboga SR, Morielle-Versute E. Morphological variation of primary reproductive structures in males of five families of neotropical bats. Anat Rec (Hoboken) 2013; 296(1): 156-67.

[http://dx.doi.org/10.1002/ar.22613] [PMID: 23117997]

[53] Beguelini MR, Góes RM, Rahal P, Morielle-Versute E, Taboga SR. Impact of the processes of testicular regression and recrudescence in the prostatic complex of the bat Myotis nigricans (Chiroptera: Vespertilionidae). J Morphol 2015; 276(7): 721-32. [http://dx.doi.org/10.1002/jmor.20373] [PMID: 25703716]

[54] Wang Z, Shi Q-X, Wang Y-N, Wang Y-L, Zhang S-Y. Epididymal sperm storage in Rickett's big-footed bat (Myotis ricketti). Acta Chiropt 2008; 10(1): 161-7. [http://dx.doi.org/10.3161/150811008X331180]

[55] León-Galván MA, Fonseca T, López-Wilchis R, Rosado A. Prolonged storage of spermatozoa in the genital tract of female Mexican big-eared bats (Corynorhinus mexicanus): the role of lipid peroxidation. Can J Zool 1999; 77(1): 7-12. [http://dx.doi.org/10.1139/z98-197]

[56] Krutzsch PH. The reproductive biology of the cave myotis (Myotis velifer). Acta Chiropt 2009; 11(1): 89-104. [http://dx.doi.org/10.3161/150811009X465712]

[57] Rodríguez-Tobón A. Análisis macro y micro estructural del epidídimo en relación con la maduración y almacenamiento de espermatozoides en el murciélago estacional Corynorhinus mexicanus. D. F., México: Universidad Autónoma Metropolitana-Iztapalapa 2011. Maestría

[58] Fouchécourt S, Métayer S, Locatelli A, Dacheux F, Dacheux JL. Stallion epididymal fluid proteome: qualitative and quantitative characterization; secretion and dynamic changes of major proteins. Biol Reprod 2000; 62(6): 1790-803. [http://dx.doi.org/10.1095/biolreprod62.6.1790] [PMID: 10819784]

[59] Belleannee C, Belghazi M, Labas V, *et al.* Purification and identification of sperm surface proteins and changes during epididymal maturation. Proteomics 2011; 11(10): 1952-64. [http://dx.doi.org/10.1002/pmic.201000662] [PMID: 21472858]

[60] Belleannée C, Labas V, Teixeira-Gomes AP, Gatti JL, Dacheux JL, Dacheux F. Identification of luminal and secreted proteins in bull epididymis. J Proteomics 2011; 74(1): 59-78. [http://dx.doi.org/10.1016/j.jprot.2010.07.013] [PMID: 20692385]

[61] Turner TT, Bomgardner D, Jacobs JP, Nguyen QA. Association of segmentation of the epididymal interstitium with segmented tubule function in rats and mice. Reproduction 2003; 125(6): 871-8. [http://dx.doi.org/10.1530/rep.0.1250871] [PMID: 12773110]

[62] Aitken RJ, Nixon B, Lin M, Koppers AJ, Lee YH, Baker MA. Proteomic changes in mammalian spermatozoa during epididymal maturation. Asian J Androl 2007; 9(4): 554-64. [http://dx.doi.org/10.1111/j.1745-7262.2007.00280.x] [PMID: 17589795]

[63] Danmaigoro A, Onu J, Sonfada M, Umaru M, Oyelowo F. Histology and Histometric Anatomy of the Male Reproductive System of Bat (Eidolon helvum). Journal of Histology 2014.

[64] Beguelini MR, Sergio BF, Leme FL, Taboga SR, Morielle-Versute E. Morphological and morphometric characteristics of the epididymis in the Neotropical bats Eumops glaucinus and Molossus molossus (Chiroptera: Molossidae). Chiroptera Neotropical 2010; 16(2): 769-79.

[65] Oliveira RL, Campolina-Silva GH, Nogueira JC, Mahecha GA, Oliveira CA. Differential expression and seasonal variation on aquaporins 1 and 9 in the male genital system of big fruit-eating bat Artibeus lituratus. Gen Comp Endocrinol 2013; 186: 116-25. [http://dx.doi.org/10.1016/j.ygcen.2013.02.041] [PMID: 23510858]

[66] Awano M, Kawaguchi A, Mohri H. Lipid composition of hamster epididymal spermatozoa. J Reprod Fertil 1993; 99(2): 375-83. [http://dx.doi.org/10.1530/jrf.0.0990375] [PMID: 8107019]

[67] Tash JS, Bracho GE. Regulation of sperm motility: emerging evidence for a major role for protein phosphatases. J Androl 1994; 15(6): 505-9.

[PMID: 7721653]

[68] Huang Y, Chung YW, Wong PY. Potassium channel activity recorded from the apical membrane of freshly isolated epithelial cells in rat caudal epididymis. Biol Reprod 1999; 60(6): 1509-14. [http://dx.doi.org/10.1095/biolreprod60.6.1509] [PMID: 10330113]

[69] Légaré C, Thabet M, Picard S, Sullivan R. Effect of vasectomy on P34H messenger ribonucleic acid expression along the human excurrent duct: a reflection on the function of the human epididymis. Biol Reprod 2001; 64(2): 720-7. [http://dx.doi.org/10.1095/biolreprod64.2.720] [PMID: 11159378]

[70] Lewis B, Aitken RJ. Impact of epididymal maturation on the tyrosine phosphorylation patterns exhibited by rat spermatozoa. Biol Reprod 2001; 64(5): 1545-56. [http://dx.doi.org/10.1095/biolreprod64.5.1545] [PMID: 11319163]

[71] Vernet P, Fulton N, Wallace C, Aitken RJ. Analysis of reactive oxygen species generating systems in rat epididymal spermatozoa. Biol Reprod 2001; 65(4): 1102-13. [http://dx.doi.org/10.1095/biolreprod65.4.1102] [PMID: 11566731]

[72] Ecroyd H, Asquith KL, Jones RC, Aitken RJ. The development of signal transduction pathways during epididymal maturation is calcium dependent. Dev Biol 2004; 268(1): 53-63. [http://dx.doi.org/10.1016/j.ydbio.2003.12.015] [PMID: 15031104]

[73] Naz RK, Rajesh PB. Role of tyrosine phosphorylation in sperm capacitation / acrosome reaction. Reprod Biol Endocrinol 2004; 2: 75. [http://dx.doi.org/10.1186/1477-7827-2-75] [PMID: 15535886]

[74] Sidhu KS, Mate KE, Gunasekera T, *et al.* A flow cytometric assay for global estimation of tyrosine phosphorylation associated with capacitation of spermatozoa from two marsupial species, the tammar wallaby (Macropus eugenii) and the brushtail possum (Trichosurus vulpecula). Reproduction 2004; 127(1): 95-103. [http://dx.doi.org/10.1530/rep.1.00073] [PMID: 15056774]

[75] Chandra A, Srinivasan KR, Jamal F, Mehrotra PK, Singh RL, Srivastav A. Post-translational modifications in glycosylation status during epididymal passage and significance in fertility of a 33 kDa glycoprotein (MEF3) of rhesus monkey (Macaca mulatta). Reproduction 2008; 135(6): 761-70. [http://dx.doi.org/10.1530/REP-07-0525] [PMID: 18502892]

[76] Baker MA, Hetherington L, Weinberg A, *et al.* Analysis of phosphopeptide changes as spermatozoa acquire functional competence in the epididymis demonstrates changes in the post-translational modification of Izumo1. J Proteome Res 2012; 11(11): 5252-64. [http://dx.doi.org/10.1021/pr300468m] [PMID: 22954305]

[77] Parks JE, Hammerstedt RH. Development changes occurring in the lipids of ram epididymal spermatozoa plasma membrane. Biol Reprod 1985; 32(3): 653-68. [http://dx.doi.org/10.1095/biolreprod32.3.653] [PMID: 3158356]

[78] Millete C. Spermatozoa. In: Knobil E, Neill JD, Eds. Encyclopedia of reproucción 2. San Diego, California: Elsevier 1999.

[79] Saez F, Frenette G, Sullivan R. Epididymosomes and prostasomes: their roles in posttesticular maturation of the sperm cells. J Androl 2003; 24(2): 149-54. [http://dx.doi.org/10.1002/j.1939-4640.2003.tb02653.x] [PMID: 12634297]

[80] Kirchhoff C, Hale G. Cell-to-cell transfer of glycosylphosphatidylinositol-anchored membrane proteins during sperm maturation. Mol Hum Reprod 1996; 2(3): 177-84. [http://dx.doi.org/10.1093/molehr/2.3.177] [PMID: 9238677]

[81] Schröter S, Osterhoff C, McArdle W, Ivell R. The glycocalyx of the sperm surface. Hum Reprod Update 1999; 5(4): 302-13. [http://dx.doi.org/10.1093/humupd/5.4.302] [PMID: 10465522]

[82] Moore A, White TW, Ensrud KM, Hamilton DW. The major maturation glycoprotein found on rat

cauda epididymal sperm surface is linked to the membrane *via* phosphatidylinositol. Biochem Biophys Res Commun 1989; 160(2): 460-8.
[http://dx.doi.org/10.1016/0006-291X(89)92455-8] [PMID: 2541702]

[83] Sullivan R, Saez F. Epididymosomes, prostasomes, and liposomes: their roles in mammalian male reproductive physiology. Reproduction 2013; 146(1): R21-35.
[http://dx.doi.org/10.1530/REP-13-0058] [PMID: 23613619]

[84] D'Amours O, Frenette G, Bordeleau LJ, *et al.* Epididymosomes transfer epididymal sperm binding protein 1 (ELSPBP1) to dead spermatozoa during epididymal transit in bovine. Biol Reprod 2012; 87(4): 94.
[http://dx.doi.org/10.1095/biolreprod.112.100990] [PMID: 22875906]

[85] Aumuller G, Wilhelm B, Seitz J. Apocrine secretion--fact or artifact? Annals of anatomy = Anatomischer Anzeiger: official organ of the Anatomische Gesellschaft 1999; 181(5): 437-46.

[86] Sullivan R, Saez F, Girouard J, Frenette G. Role of exosomes in sperm maturation during the transit along the male reproductive tract. Blood Cells Mol Dis 2005; 35(1): 1-10.
[http://dx.doi.org/10.1016/j.bcmd.2005.03.005] [PMID: 15893944]

[87] Rejraji H, Sion B, Prensier G, *et al.* Lipid remodeling of murine epididymosomes and spermatozoa during epididymal maturation. Biol Reprod 2006; 74(6): 1104-13.
[http://dx.doi.org/10.1095/biolreprod.105.049304] [PMID: 16510839]

[88] Sostaric E, Aalberts M, Gadella BM, Stout TA. The roles of the epididymis and prostasomes in the attainment of fertilizing capacity by stallion sperm. Anim Reprod Sci 2008; 107(3-4): 237-48.
[http://dx.doi.org/10.1016/j.anireprosci.2008.04.011] [PMID: 18550301]

[89] Frenette G, Légaré C, Saez F, Sullivan R. Macrophage migration inhibitory factor in the human epididymis and semen. Mol Hum Reprod 2005; 11(8): 575-82.
[http://dx.doi.org/10.1093/molehr/gah197] [PMID: 16051683]

[90] Frenette G, Girouard J, Sullivan R. Comparison between epididymosomes collected in the intraluminal compartment of the bovine caput and cauda epididymidis. Biol Reprod 2006; 75(6): 885-90.
[http://dx.doi.org/10.1095/biolreprod.106.054692] [PMID: 16943362]

[91] Frenette G, Girouard J, D'Amours O, Allard N, Tessier L, Sullivan R. Characterization of two distinct populations of epididymosomes collected in the intraluminal compartment of the bovine cauda epididymis. Biol Reprod 2010; 83(3): 473-80.
[http://dx.doi.org/10.1095/biolreprod.109.082438] [PMID: 20554923]

[92] Toshimori K. Maturation of mammalian spermatozoa: modifications of the acrosome and plasma membrane leading to fertilization. Cell Tissue Res 1998; 293(2): 177-87.
[http://dx.doi.org/10.1007/s004410051110] [PMID: 9662641]

[93] Eddy EM, Vernon RB, Muller CH, Hahnel AC, Fenderson BA. Immunodissection of sperm surface modifications during epididymal maturation. Am J Anat 1985; 174(3): 225-37.
[http://dx.doi.org/10.1002/aja.1001740305] [PMID: 3907324]

[94] Bearer EL, Friend DS. Morphology of mammalian sperm membranes during differentiation, maturation, and capacitation. J Electron Microsc Tech 1990; 16(4): 281-97.
[http://dx.doi.org/10.1002/jemt.1060160403] [PMID: 2250184]

[95] Moore H. The influence of the epididymis on human and animal sperm maturation and storage. Hum Reprod 1996; 11: 103-10.
[PMID: 9147105]

[96] Tulsiani DR. Glycan-modifying enzymes in luminal fluid of the mammalian epididymis: an overview of their potential role in sperm maturation. Mol Cell Endocrinol 2006; 250(1-2): 58-65.
[http://dx.doi.org/10.1016/j.mce.2005.12.025] [PMID: 16413674]

[97] Toyonaga M, Morita M, Hori T, Tsutsui T. Distribution of glycoproteins on feline testicular sperm, epididymal sperm and ejaculated sperm. The Journal of veterinary medical science / the Japanese

Society of Veterinary Science 2011; 73(6): 827-9.
[http://dx.doi.org/10.1292/jvms.10-0400]

[98] Baker MA, Nixon B, Naumovski N, Aitken RJ. Proteomic insights into the maturation and capacitation of mammalian spermatozoa. Syst Biol Reprod Med 2012; 58(4): 211-7.
[http://dx.doi.org/10.3109/19396368.2011.639844] [PMID: 22788533]

[99] Yanagimachi R. Mammalian Fertilization. In: Knobil E, Neill D, Eds. The Physiology of Reproduction. New York: Raven Press 1988; pp. 135-85.

[100] Agarwal A, Said TM. Carnitines and male infertility. Reprod Biomed Online 2004; 8(4): 376-84.
[http://dx.doi.org/10.1016/S1472-6483(10)60920-0] [PMID: 15149558]

[101] Ahuja KK. Carbohydrate determinants involved in mammalian fertilization. Am J Anat 1985; 174(3): 207-23.
[http://dx.doi.org/10.1002/aja.1001740304] [PMID: 4072940]

[102] Holt WV. Surface-bound sialic acid on ram and bull spermatozoa: deposition during epididymal transit and stability during washing. Biol Reprod 1980; 23(4): 847-57.
[http://dx.doi.org/10.1095/biolreprod23.4.847] [PMID: 7448285]

[103] Bedford JM. Morphological changes in rabbit spermatozoa during passage through the epididymis. J Reprod Fertil 1963; 5: 169-77.
[http://dx.doi.org/10.1530/jrf.0.0050169] [PMID: 13970203]

[104] Moore HD, Bedford JM. Short-term effects of androgen withdrawal on the structure of different epithelial cells in the rat epididymis. Anat Rec 1979; 193(2): 293-311.
[http://dx.doi.org/10.1002/ar.1091930209] [PMID: 426300]

[105] Arenas MI, de Miguel MP, Bethencourt FR, Fraile B, Royuela M, Paniagua R. Lectin histochemistry in the human epididymis. J Reprod Fertil 1996; 106(2): 313-20.
[http://dx.doi.org/10.1530/jrf.0.1060313] [PMID: 8699416]

[106] Fierro R, Foliguet B, Grignon G, *et al.* Lectin-binding sites on human sperm during acrosome reaction: modifications judged by electron microscopy/flow cytometry. Arch Androl 1996; 36(3): 187-96.
[http://dx.doi.org/10.3109/01485019608987095] [PMID: 8743350]

[107] Vazquez JM, Martinez E, Pastor LM, Roca J, Matas C, Calvo A. Lectin histochemistry during *in vitro* capacitation and acrosome reaction in boar spermatozoa: new lectins for evaluating acrosomal status of boar spermatozoa. Acta Histochem 1996; 98(1): 93-100.
[http://dx.doi.org/10.1016/S0065-1281(96)80056-0] [PMID: 9054195]

[108] Flesch FM, Voorhout WF, Colenbrander B, van Golde LM, Gadella BM. Use of lectins to characterize plasma membrane preparations from boar spermatozoa: a novel technique for monitoring membrane purity and quantity. Biol Reprod 1998; 59(6): 1530-9.
[http://dx.doi.org/10.1095/biolreprod59.6.1530] [PMID: 9828202]

[109] Rodríguez-Tobón A. Estudio de la maduración de espermatozoides epididimarios del murciélago estacional Corynorhinus mexicanus [Doctorado]. Ciudad de México: Universidad Autónoma metropolitana 2016.

[110] Robaire B, Seenundun S, Hamzeh M, Lamour SA. Androgenic regulation of novel genes in the epididymis. Asian J Androl 2007; 9(4): 545-53.
[http://dx.doi.org/10.1111/j.1745-7262.2007.00316.x] [PMID: 17589794]

[111] Hess RA. Estrogen in the adult male reproductive tract: a review. Reprod Biol Endocrinol 2003; 1(1): 52.
[http://dx.doi.org/10.1186/1477-7827-1-52] [PMID: 12904263]

[112] Patrão MT, Silva EJ, Avellar MC. Androgens and the male reproductive tract: an overview of classical roles and current perspectives. Arq Bras Endocrinol Metabol 2009; 53(8): 934-45.
[http://dx.doi.org/10.1590/S0004-27302009000800006] [PMID: 20126845]

[113] Hinton BT, Lan ZJ, Lye RJ, Labus JC. Regulation of epididymal function by testicular factors: the lumicrine hypothesis The testis. Springer 2000; pp. 163-73.

[114] Sairam MR, Krishnamurthy H. The role of follicle-stimulating hormone in spermatogenesis: lessons from knockout animal models. Arch Med Res 2001; 32(6): 601-8. [http://dx.doi.org/10.1016/S0188-4409(01)00328-9] [PMID: 11750736]

[115] Turrens JF. Mitochondrial formation of reactive oxygen species. J Physiol 2003; 552(Pt 2): 335-44. [http://dx.doi.org/10.1113/jphysiol.2003.049478] [PMID: 14561818]

[116] Chabory E, Damon C, Lenoir A, *et al.* Mammalian glutathione peroxidases control acquisition and maintenance of spermatozoa integrity. J Anim Sci 2010; 88(4): 1321-31. [http://dx.doi.org/10.2527/jas.2009-2583] [PMID: 20042549]

[117] Arenas-Ríos E, Rosado A, Rodríguez-Tobón E, Rodríguez-Tobón A, León-Galván MA. Gamma glutamyl transpeptidase activity determination in epididymis of Corynorhinus mexicanus bat throughout Its annual cycle. Animal and Veterinary Sciences Special Issue: Advances in Bat's Reproduction 2015; 3(1-1): 8-12.

[118] Aitken RJ, Paterson M, Fisher H, Buckingham DW, van Duin M. Redox regulation of tyrosine phosphorylation in human spermatozoa and its role in the control of human sperm function. J Cell Sci 1995; 108(Pt 5): 2017-25. [PMID: 7544800]

[119] Aitken RJ, Gordon E, Harkiss D, *et al.* Relative impact of oxidative stress on the functional competence and genomic integrity of human spermatozoa. Biol Reprod 1998; 59(5): 1037-46. [http://dx.doi.org/10.1095/biolreprod59.5.1037] [PMID: 9780307]

[120] de Lamirande E, Jiang H, Zini A, Kodama H, Gagnon C. Reactive oxygen species and sperm physiology. Rev Reprod 1997; 2(1): 48-54. [http://dx.doi.org/10.1530/ror.0.0020048] [PMID: 9414465]

[121] Leclerc P, de Lamirande E, Gagnon C. Regulation of protein-tyrosine phosphorylation and human sperm capacitation by reactive oxygen derivatives. Free Radic Biol Med 1997; 22(4): 643-56. [http://dx.doi.org/10.1016/S0891-5849(96)00379-6] [PMID: 9013127]

[122] Baker MA, Aitken RJ. The importance of redox regulated pathways in sperm cell biology. Mol Cell Endocrinol 2004; 216(1-2): 47-54. [http://dx.doi.org/10.1016/j.mce.2003.10.068] [PMID: 15109744]

[123] Caselli A, Marzocchini R, Camici G, *et al.* The inactivation mechanism of low molecular weight phosphotyrosine-protein phosphatase by H2O2. J Biol Chem 1998; 273(49): 32554-60. [http://dx.doi.org/10.1074/jbc.273.49.32554] [PMID: 9829991]

[124] Aitken RJ. Possible redox regulation of sperm motility activation. J Androl 2000; 21(4): 491-6. [PMID: 10901434]

[125] Ford WC. Regulation of sperm function by reactive oxygen species. Hum Reprod Update 2004; 10(5): 387-99. [http://dx.doi.org/10.1093/humupd/dmh034] [PMID: 15218008]

[126] Gamou S, Shimizu N. Hydrogen peroxide preferentially enhances the tyrosine phosphorylation of epidermal growth factor receptor. FEBS Lett 1995; 357(2): 161-4. [http://dx.doi.org/10.1016/0014-5793(94)01335-X] [PMID: 7805884]

[127] Guyton KZ, Liu Y, Gorospe M, Xu Q, Holbrook NJ. Activation of mitogen-activated protein kinase by H2O2. Role in cell survival following oxidant injury. J Biol Chem 1996; 271(8): 4138-42. [http://dx.doi.org/10.1074/jbc.271.8.4138] [PMID: 8626753]

[128] Tonks NK. PTP1B: from the sidelines to the front lines! FEBS Lett 2003; 546(1): 140-8. [http://dx.doi.org/10.1016/S0014-5793(03)00603-3] [PMID: 12829250]

[129] van Montfort RL, Congreve M, Tisi D, Carr R, Jhoti H. Oxidation state of the active-site cysteine in

protein tyrosine phosphatase 1B. Nature 2003; 423(6941): 773-7. [http://dx.doi.org/10.1038/nature01681] [PMID: 12802339]

[130] Zipser Y, Piade A, Kosower NS. Erythrocyte thiol status regulates band 3 phosphotyrosine level *via* oxidation/reduction of band 3-associated phosphotyrosine phosphatase. FEBS Lett 1997; 406(1-2): 126-30. [http://dx.doi.org/10.1016/S0014-5793(97)00263-9] [PMID: 9109401]

[131] Shalgi R, Seligman J, Kosower NS. Dynamics of the thiol status of rat spermatozoa during maturation: analysis with the fluorescent labeling agent monobromobimane. Biol Reprod 1989; 40(5): 1037-45. [http://dx.doi.org/10.1095/biolreprod40.5.1037] [PMID: 2765609]

[132] López-Wilchis R. Biología de Plecotus mexicanus (Chiroptera: Vespertilionidae) en el estado de Tlaxcala, México [Doctoral thesis]. México DF: Universidad Nacional Autónoma de México 1989.

[133] Arenas-Ríos E. Enzimas anti-especies reactivas de oxígeno, como reguladores en los procesos de espermatogénesis maduración y almacenamiento prolongado de espermatozoides en el murciélago Corynorhinus mexicanus. México: Universidad Autónoma Metropolitana 2009. Doctorado

CHAPTER 6

Mechanisms of Sperm-Storage in Female Reproductive Tract in Bats

Amitabh Krishna[1,*] and **Vikas Kumar Roy**[1,2]

[1] *Department of Zoology, Institute of Sciences, Banaras Hindu University, Varanasi-221005, India*

[2] *Department of Zoology, School of Life Sciences, Mizoram University, Aizawl-796004, India*

Abstract: Prolonged sperm-storage is an adaptation to the reproductive cycle that is most highly developed in bat. Twenty four bat species have been identified so far to store spermatozoa for the period of approx. 16 days up to ~225 days and still retain fertilizing capacity. The aim of this review is to evaluate the mechanism and molecules necessary for sperm to become efficiently stored in female genital tract. Spermatozoa are arranged with their head towards the reproductive tract epithelium are characterized in all sperm storing bat species so far investigated, suggesting that such intimate relationships are an integral part of the mechanism of prolonged storage of sperm. Recent study suggests that the ultimate controls of the mechanism underlying sperm-storage are hormonal. It has been demonstrated that sperm stored in female bat contains high circulating level of androgen during the period of sperm storage. It is suggested that androgen creates a unique microenvironment that facilitates prolonged sperm storage. The identification of the specific molecules responsible for prolonged sperm-storage may suggest a mechanism to keep the sperm viable for prolonged period for use in assisted reproductive technique.

Keywords: Androgen, Apoptosis, Bat, Female, Sperm, Storage.

INTRODUCTION

The phenomenon of sperm storage in female reproductive tract has been evolved in several species which belong to different taxonomic groups, including, insects, fishes; reptiles, birds, and mammals [1 - 10]. Among mammals, prolonged sperm storage is an adaptation to the reproductive cycle that is most highly developed in hibernating chiropteran from temperate latitude [11 - 14]. Ability of female bat to store sperm is an integral part of the reproductive strategy and can provide important advantages to the species. Sperm storage typically occurs in females, whose folliculogenesis or formation of pre-ovulatory follicle is not synchronized

* **Corresponding author Amitabh Krishna:** Reproductive Physiology Unit, Department of Zoology, Institute of Science, Banaras Hindu University, Varanasi-221 005, India; Tel: 0542-6702527; E-mail: akrishna_ak@yahoo.co.in

Edith Arenas-Ríos (Ed.)

with that of the male or in males that can retain spermatozoa beyond the cessation of spermatogenesis. Although, sperm-storage is an important feature of reproductive cycle in the majority of vespertilionid and rhinolophid bats, a similar phenomenon has also been recorded in temperate zone molossid bat, *Mormopterus planiceps* [15]. In temperate zone bat, mating activity commences in the autumn, spermatozoa are stored in the female genital tract throughout the hibernating period. These sperms are used in the fertilization of ova released at ovulation after arousal in spring [12, 16]. The period of sperm-storage in the female genital tract in bat ranges from 16 days as in *Pipistrellus ceylonicus* [17] to ~198 days as in *Nyctalus noctula* [18]. Other mammal species (*e.g.* dog [19]; hare [20], showed sperm-storage for relatively short periods. Tropical bats generally store sperm for much shorter period as compared with hibernating temperate zone bats. The interval between copulation and ovulation is up to five weeks in *Scotophilus heathi* [21] and 30-35 days in *Pipistrellus mimus* [22]. The nature of specialization that has evolved in the bats to permit sperm survival for prolonged period is derived mainly from the studied on some vespertilionid bats. In majority of female bats, sperms are stored in specialized organs or reservoirs, where sperms are retained for variable duration. This review aims to evaluate the site of sperm-storage and molecules necessary for sperm to become efficiently stored for prolonged period in female bats.

SPERM-STORAGE SITE

Sperm storage in female genital tract normally exhibits in specialized storage region in different vertebrate species. In some sharks, reptiles and birds, storage of sperm largely observed within specialized blind ended tubules lined with epithelial cells known as sperm-storage tubules [23]. In these animals, the sperm-storage tubules are located at utero-vaginal junction of female genital tract. The caecilian amphibian possess spermathecae for storage. Mammalian groups such as marsupial and insectivore have developed sperm-storage crypts [24]. Spermatozoa are usually arranged within their heads pointing toward epithelial lining of tubule until the time of its release. Sperm motility is suppressed during storage and restored at the time of release. The sperm-storage tubules are rich in antioxidants that protect sperm against harmful oxidant and inappropriate immune threat.

Site of sperm storage in the female genital tract varies from peri-ovarian spaces [25] to the oviductal isthmus, utero-tubal junction (UTJ), uterus, cervix and vagina, depending on the species [11]. In some vespertilionid bats, viz. *Chalinolobus gouldii* [26] and *M. schreibersii fuliginosus*, the storage site is limited to the UTJ where intact spermatozoa are found in diverticulae formed by the mucosal folds [27]. The sperms are stored in UTJ in *M. daubentoni* [11], *Myotis lucifugus* and *Myotis velifer* [28]. In *P. ceylonicus* and *S. heathi*, surviving spermatozoa are stored mainly in UTJ as well as in both the uterus, and oviduct (Fig. **1**) [7, 21]. In *P. mimus*, comparatively few spermatozoa are found in the uterus and most occur in the oviduct. In *Rhinolophus ferrum-equinum*, *R. hipposideros, R. ferrum equinum Nippon* and *Myotis nattereri*, the principal site of sperm-storage is the oviduct [11, 28]. The storage organs are thought to provide nourishment and/or protection for the sperm, allowing them to remain viable for longer periods [29, 30]. Earlier studies showed a close association between sperm and the epithelial lining of the storage organs is observed in chickens, bats, boars, hamsters, and other mammals [29, 31]. The epithelial cells of sperm-storage organs (tubules) are rich in lipids, β-glycogen, mitochondria, and secretary granules [32], suggesting that these cells produce secretions utilized by the sperm. Cells in the female storage organ may provide nutrients or metabolites to sperm to maintain viability or encourage quiescence during prolonged storage.

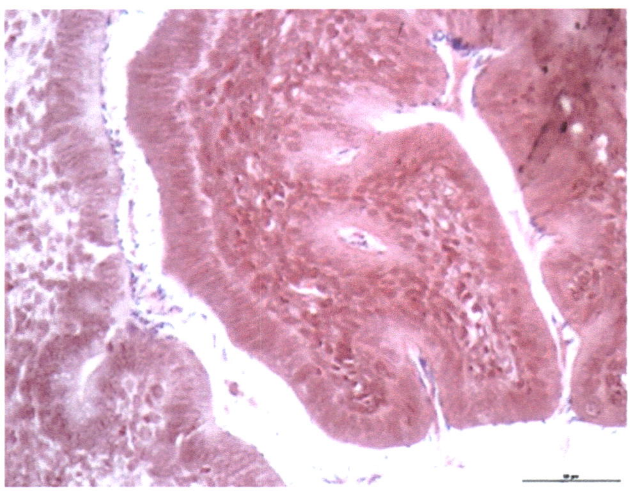

Fig. (1). Histological section of female reproductive tract of *S. heathi* section during sperm storage period showing many spermatozoa arranging their heads towards the uterine epithelium.

MECHANISM OF SPERM-STORAGE

The mechanism of sperm-storage has been evolved several times [33]. Whether different taxonomic groups adapt to same mechanism is not clear from the data available so far. Studies have been conducted largely in bats to understand the biology of prolonged sperm longevity by morphological [11, 30, 34-37; metabolic [38 - 41]; experimental, hormonal and nutritional approaches [42, 43]. Although, several theories have been offered to explain sperm-storage, however, none of the underlying mechanism has been unequivocally substantiated. Recent studies on *Scotophilus heathi* provided some novel concepts about prolonged sperm storage process.

Low Body Temperature Facilitate Sperm-storage

Prolonged storage of viable spermatozoa within the female genital tract was known earlier in various poikilothermic animals [44], but among mammal, the phenomenon is more extensively investigated in bats mainly due to their heterothermic characteristic. Pagenstecher [45] first drew attention to the fact that spermatozoa are stored in the uterus of the Pipistrellus bat, *Pipistrellus pipistrellus* during hibernation. Subsequent reports of sperm-storage in the reproductive tract of both male and female vespertilionid and rhinolophid bats were mainly from temperate regions. In the hibernating bats from Europe, North America and Japan, there is a long interval between copulation and ovulation and fertilization, and during this period, sperm remain stored, which coincides with the period of hibernation [16, 46]. Such studies suggest a direct functional relationship between sperm-storage and winter dormancy [47]. Winter dormancy was also considered to be important for sperm-storage reducing loss or degradation of spermatozoa [18], and inhibiting ovulation [47]. Hartman [48] suggested that low body temperature of the hibernating bat reduces sperm motility and thereby favors their survival. Discovery of sperm-storage phenomenon in the tropical vespertilionids, *Tylonycteris pachypus* and *T. robustula* [49], *Pipistrellus ceylonicus* [17] and *Scotophilus heathi* [21], suggests that [21], suggests that which characterizes temperate zone species, is not a necessary preadaptation for sperm-storage.

Unique Relationship between Spermatozoa and Storage Organ

In most chiropterans showing sperm-storage for the prolonged period, spermatozoa are usually perpendicularly arranged with their heads pointing towards the epithelium cells of reproductive tract [11, 22]. Such intimate relationship exists between spermatozoa and female storage organs in all sperm storing bat species and are an integral part of the mechanism of prolonged sperm-storage [12, 43]. This relationship is more pronounced in the area around the entrance of UTJ [50]. Similar intimate relationship between the head of spermatozoa and storage organs has also been described in other vertebrates that retain sperm for prolonged period (*e.g.* garter snake [51]: birds [52]; dog [19]. During storage, the sperm heads are either embedded in the uterine and oviductal epithelia or remain in close contact with microvilli of the host cells. Some spermatozoa actually indenting the epithelial cells to the point that their heads are enveloped in epithelial cells cytoplasm (*e.g. Pipistrellus kuhli* [53]; *Myotis lucifugus* and *M. velifer* [11, 36]. Such a relationship was shown in *Pipistrllus kuhli* to resist trypsin digestion but succumb to treatment with EDTA and hyaluronidase, suggesting that attachment is mediated by proteoglycans or Ca++ rather than a protein [53]. According to Racey [11], the epithelial cells of the storage site actively provide spermatozoa with nutrients for their prolonged survival. For example, glycogen in the uterine epithelium of *P. abramus* [54], *P. pipistrellus* [11], *Myotis lucifugus* and *M. velifer* [38].

It was further suggested that the association of spermatozoa with the epithelium in the storage organ probably protects spermatozoa from various disintegrative factors (*e.g.* muscular movements of the uterus). Certain enzymes of glycolysis have been located in the sperm as well as in the uterus [11]. Beside nutritional support, other factors required for long term maintenance of spermatozoa might be a mechanism that acts to suppress sperm metabolism, thereby reducing demand for available substrates. In this regard, Hunter *et al.* [55, 56], observed that the seminal fluid of *M. lucifugus* contains a protein that is inhibitory to uterine motility and phagocytic system. Zinc has been shown to temporarily inhibit the motility of both sperm and polymorphonuclear neutrophilic leucocytes. It has been suggested that such factors would also favor prolonged sperm storage [39]. Oxidative damage is a potential problem for stored spermatozoa in female genital tract [23].

Hormonal Regulation of Sperm-Storage

It appears that the ultimate control of the mechanism underlying sperm-storage may be hormonal. Racey [11] ovariectomized *Pipistrellus pipistrellus* during hibernation, causing detrimental effects on the survival of stored sperm after two weeks as compared with control bats. Circulating levels of androgens that are required for the maintenance of sperm integrity and viability in the male are known for only a few bat species. Circulating androstenedione may be present in elevated levels in females storing sperm [57]. Androgen receptors (ARs) and androgen binding proteins (ABPs) are present in the epithelial cell lining of the utero-tubal junction (storage organ) during the period of sperm-storage in *Scotophilus heathi* [58]. The results of recent study further confirm the role of androgen in sperm-storage. The female bat treated with anti-androgen (Flutamide) during the period of sperm-storage caused disappearance or loss of storing spermatozoa together with significant decline in androgen level. This suggests that the decline in androgen activates changes in the epithelial cells that may promote degradation and disposal of spermatozoa [59]. Recent studies have suggested that sperm release from the storage sites correlates with increased progesterone release during peri-ovulatory period [23]. Recent immunocytochemical study revealed the presence of increased expression of BCL2 in the epithelial lining of utero-tubal junction of *S. heathi* during the period of sperm-storage. Treatment with testosterone *in vitro* caused dose-dependent significant increase in BCL2 expression, whereas treatment with flutamide together with testosterone caused significant decrease in BCL2 expression in utero-tubal junction of *S. heathi*. The utero-tubal junction also expresses caspase 3 together with BCL2. The expression of caspase 3 decreased significantly during January, the period of sperm-storage, but increases during February prior to fertilization. This study provides evidence suggesting that androgen promotes sperm-storage by stimulating BCL2 synthesis in the site of sperm-storage. It may be hypothesized that a balance between BCL2 and caspase 3 selects the healthier sperm for survival and is subsequently utilized for fertilization.

Nutritional Support and Metabolic Changes

Recent studies from our laboratory and earlier study of Racey [11] suggested the role of glucose and fatty acid as nutrient substances to the storing sperm and provided evidence of the transfer of these metabolites from uterine cells to stored spermatozoa. Racey [11] provided a comprehensive light microscopical report on the histochemical analysis of the presence of lipids (phospholipids), proteins and carbohydrates in sperm storing tissues of both male and female bat, *Pipistrellus pipistrellus*. Subsequent studies at the electron microscope level have confirmed the presence of glycogen in sperm-storage site (oviduct, utero-tubal junction) [30, 36, 53] giving rise to speculation that this may be an energy source for stored sperm. Enzymes of glycolysis have also been located in the sperm as well as in the uterus [11]. Recent study from our laboratory showed the presence of GLUT 3, 4, 5 and 8 in the UTJ of the bat, *S. heathi* [60]. The significant decrease in serum glucose level together with decrease in GLUT 3 and 5 proteins in the UTJ suggests decreased glucose availability during the period of sperm-storage. This study suggests that the decline in the availability of glucose to the stored sperm in UTJ may be responsible for lowered sperm motility which consequently promotes sperm storage in *S. heathi* [60]. Suppressed sperm motility has been suggested as a common feature of natural sperm-storage process [33]. Increased GLUT mediated uptake of glucose and production of energy through glucose metabolism are required for fertilizing ability of sperm, and decline in glucose availability as found in diabetes mellitus is shown to impair fertility [61, 62].

During glucose deficiency, spermatozoa primarily utilize lipids as a source of energy to store sperms. Further study showed significant increased level of hormone sensitive lipase and carnitine in the utero-tubal junction suggests increased fatty acid availability during the period of sperm storage, which may be utilized by sperm during the storage period. Increased carnitine also maintains integrity of sperm during storage. Besides nutritional support, suppress sperm metabolism may be another important factor required for long-term maintenance of sperm, thereby reducing demand for available substrates. Hunter *et al*. [55, 56], observed that the seminal fluid of *M. lucifugus* contains a protein that is inhibitory to uterine sperm motility. The metallic ion Zn is one such substance to inhibit the activity of sperm.

Capacitation and Sperm-Storage

Krutzsch & Crichton [63] suggested that bat spermatozoa resist capacitation and the acrosome reaction when cultured under conditions that are known to induce these events in other mammals. Furthermore, they have failed to reveal decondensing bat sperm heads within zona-free hamster ooplasm using *in vitro* fertilization procedure, even after exposure to calcium ionophore A23187, and various proteases. Thus, it was suggested that the unusual longevity, stability and resistance of bat spermatozoa to *in vitro* fertilization could be due to unique modification of their plasma membrane. A recent study on a hibernating bat (*Myotis velifer*), however, has failed to demonstrate any modification in plasma membrane lipid composition of the stored spermatozoa [40]. Further studies are needed to find out the unique features of prolonged sperm longevity of bats.

It has been suggested that prolonged sperm survival in female tract including bats might involve a delay in capacitation, because capacitated sperm can be released from binding site and could be phagocytized soon [36, 64 - 66]. Capacitation is a complex molecular process that takes place in sperm when comes in contact with the female reproductive tract and is defined as the physiological events that render the sperm able to fertilize [67, 68]. Once sperms are capacitated and undergo acrosome reaction, they do not survive longer [64]. If the female tract is dependent on the elevated level of androgen, then it might affect capacitation, sperm survival and sperm binding and release to and from female tract epithelium. As suggested earlier, it has been hypothesized that release of the sperm from the oviductal epithelium is due to their capacitated state [29]. Moein-Vaziri *et al.* [69], demonstrated that a brief exposure of boar spermatozoa to extracellular recombinant heat shock protein (HSP)-A8 significantly increased sperm viability, decreased mitochondrial activity, and increased the capacity of sperm to bind oviductal epithelium cells. The exact role of HSPs has not been studied in relation to sperm storage in the female tract.

Studies on Bats, *Scotophilus heathi*:

The studies undertaken primarily on *S. heathi* provided sufficient evidences to understand the basic mechanism of sperm storage in bat in general. This species displays a clear dissociation between the timing of male and female gamete production. Mating in *S. heathi*, occurs in early winter and spermatozoa remain stored mainly in UTJ until ovulation in late February or early March [21]. The bat accumulates white adipose tissue prior to winter, resulting in one and half time increase in the body mass [70]. Increase in body mass occurs as a consequence of overfeeding which leads to increase in insulin release. Hyperinsulinemia promotes increased androgen production from thecal cells of antral or cystic follicles. Subsequently a feature resembles-like polycystic ovary syndrome (PCOS) is formed and ovulation is delayed in *S. heathi* [71]. Together with the increased androgen production, the oviductal epithelia shows the presence of androgen receptors and androgen binding protein (Fig. **2A**, **B**) [58].

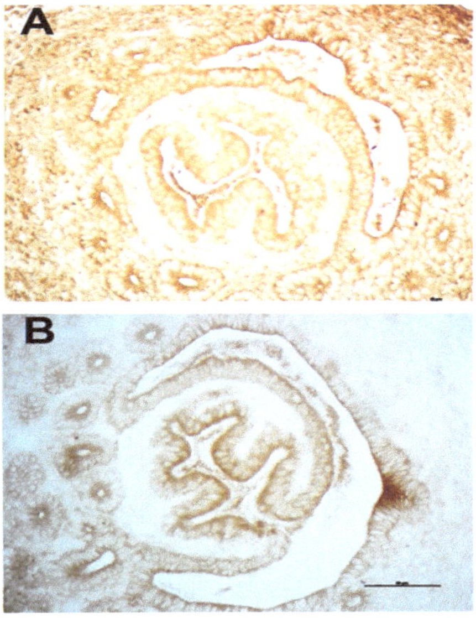

Fig. (2). Immunolocalization of androgen binding protein (A) and androgen receptor (B) in the female reproductive tract of *S. heathi* during sperm storage. Both androgen binding protein and androgen receptor show cytoplasmic immunostaining in the luminal epithelial cells of utero-tubal junction.

The ovary of the bat produced increases androgen on stimulation by increased circulating insulin level during the period of increased fat accumulation and is accompanied by a decline in circulating glucose and carnitine levels [60]. The expression of glucose transporters (GLUT 3 and GLUT 5) declines, whereas the carnitine concentration, the expression of carnitine transporter (OCTN2) and hormone sensitive lipase increase at UTJ during the period of sperm- storage [60]. These findings suggest that decreased blood glucose concentration but increased level of free fatty acid provide condition suitable for sperm storage in *S. heathi*. Extensive studies on sperm-storage conducted on *S. heathi* emphasized that the entire physiology such as hypothalamic-pituitary-ovarian axis, feeding regulation, control of hibernation, role of adipose tissue may also be involve in this unique phenomena of prolonged sperm-storage [23]. The mechanism of sperm-storage proposed based on extensive observations obtained from the studies conducted on a single species of bat, *S. heathi*, requires further confirmation on other species

(Fig. **3**). The process of sperm storage has recently been investigated in the female reproductive tract of the Chinese soft-shelled turtle, *Pelodiscus sinensis* [72, 73]. These studies have also shown involvement of androgen and BCL2 in sperm storage in *P. sinensis* [73].

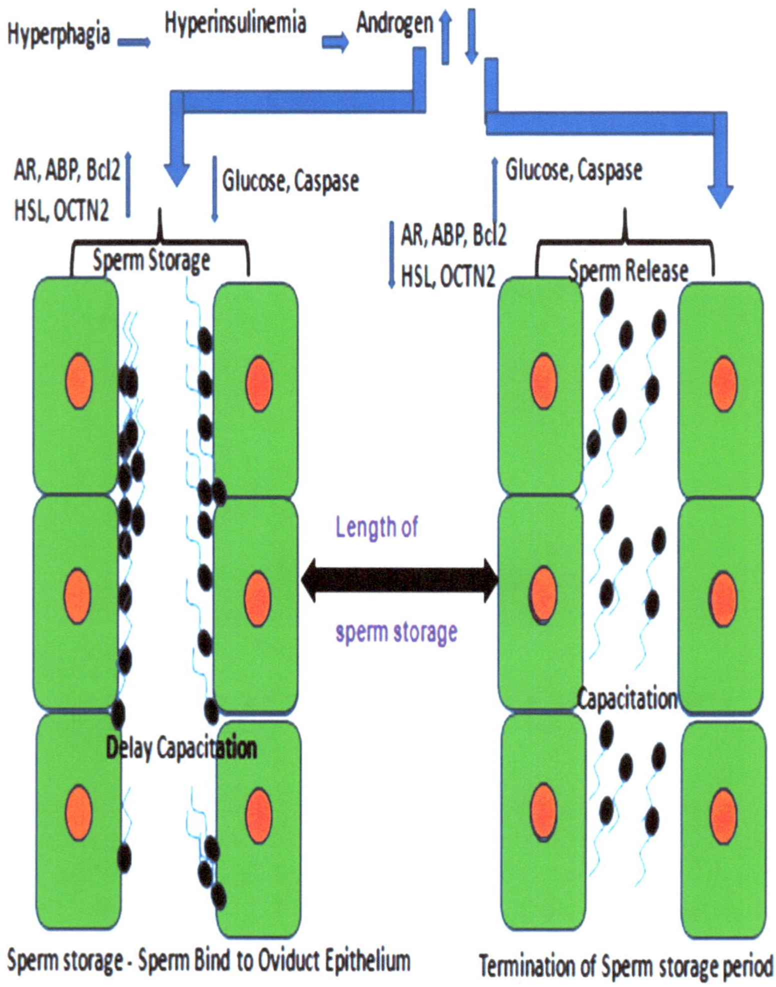

Fig. (3). A schematic diagram of possible mechanisms of sperm storage in the female reproductive tract of *S. heathi*. This diagram shows high androgen concentration along with elevated levels of androgen receptor (AR) and androgen binding protein (ABP) promotes sperm storage by delaying capacitation and preventing sperm loss from death *via* increasing expression of Bcl2. The sperm storage period coincides with low glucose which further facilitates sperm storage and sorted fat provides energy for stored sperm by regulating hormone sensitive lipase (HSL) and carnitine transporter (OCTN2). The low concentration of androgen terminates sperm storage and release sperm from storage site by promoting capacitation.

SIGNIFICANCE OF SPERM STORAGE

The mechanism of sperm-storage may be utilized to develop successful reproduction. Questions arise, would sperm-storage process be simulated for the development of novel approaches for long term sperm-storage *in vitro*? In a recent review by Holt & Fazeli [23], strongly advocated idea of long term sperm storage *in vitro* for an alternative to cryopreservation. Recent studies suggest that the female reproductive tract of sperm-storing bat provides a specialized environment that interacts with the spermatozoa *via* hormonally-mediated pathways, and protecting against harmful reactive antioxidants and immune threats. A detailed biochemical and molecular knowledge about the sperm-storage site (UTJ) of the bat may help in developing successful long-term method for prolonged sperm survival *in vitro* for use in reproductive technologies. Ashizawa and Colleagues [74 - 78] showed that sperm surviving *in vitro* was improved significantly by co-culture with a variety of tissues associated with sperm-storage. Thus it would be interesting to try to co-culture sperm with UTJ of hyperandrogenised animal to simulate the long term sperm-storage in *in vitro* condition.

CONCLUSION

Storage of sperm is an important strategy of female reproductive cycle of bat. Normally in mammals that don't store sperm, large numbers of white blood cells gather and phagocytize the sperm cells. Therefore, sperm-storage is a remarkable process because "foreign" cells, which are normally destroyed by the immune system, are not only tolerated, but perhaps even nourished for several months in the female [79]. It appears that sperm storage may be possible by activity that tends to prevent both fertilization and premature cell death. Demonstration of increased secretion of BCL2 (survival factor) by utero-tubal junction suggests its role in sperm viability. Decreased sperm motility due to insufficient availability of nutrient (*e.g.* glucose, lipids) may be a possible feature of sperm storage mechanism and need further investigation. We have not discussed the likelihood that sperm DNA-repair mechanisms might also be involved. Presence of heat shock protein on stored spermatozoa of boar [80], thus its role in sperm viability requires further investigation.

ABBREVIATIONS

AR androgen receptor

ABP androgen binding protein

BCL2 B-cell lymphoma factor 2

HSL hormone sensitive lipase

OCTN2 organic cation transporter 2

UTJ utero-tubal junctions

CONFLICT OF INTEREST

The authors declare no conflict of interest, financial or otherwise.

ACKNOWLEDGEMENTS

AK highly acknowledges the financial assistance from the Department of Science & Technology, New Delhi and Centre of Advance Study, University Grant Commission, New Delhi.

REFERENCES

[1] Birkhead TR, Moller AP. Sexual selection and the temporal separation of reproductive events: sperm storage data from reptiles, birds and mammals. Biol J Linn Soc Lond 1993; 50: 295-311. [http://dx.doi.org/10.1111/j.1095-8312.1993.tb00933.x]

[2] Gist DH, Congdon JD. Oviductal sperm storage as a reproductive tactic of turtles. J Exp Zool 1998; 282(4-5): 526-34. [http://dx.doi.org/10.1002/(SICI)1097-010X(199811/12)282:4/5<526::AID-JEZ8>3.0.CO;2-Z] [PMID: 9803538]

[3] Neubaum DM, Wolfner MF. Wise, winsome, or weird? Mechanisms of sperm storage in female animals. Curr Top Dev Biol 1999; 41: 67-97. [http://dx.doi.org/10.1016/S0070-2153(08)60270-7] [PMID: 9784973]

[4] Yamanouye N, Silveira PF, Abdalla FM, Almeida-Santos SM, Breno MC, Salomão MG. Reproductive cycle of the Neotropical *Crotalus durissus terrificus*: II. Establishment and maintenance of the uterine muscular twisting, a strategy for long-term sperm storage. Gen Comp Endocrinol 2004; 139(2): 151-7. [http://dx.doi.org/10.1016/j.ygcen.2004.09.002] [PMID: 15504393]

[5] Baer B, Armitage SA, Boomsma JJ. Sperm storage induces an immunity cost in ants. Nature 2006; 441(7095): 872-5. [http://dx.doi.org/10.1038/nature04698] [PMID: 16778889]

[6] Das SC, Isobe N, Nishibori M, Yoshimura Y. Expression of transforming growth factor-β isoforms and their receptors in utero-vaginal junction of hen oviduct in presence or absence of resident sperm with reference to sperm storage. Reproduction 2006; 132(5): 781-90. [http://dx.doi.org/10.1530/rep.1.01177] [PMID: 17071779]

[7] Zhang YY. Semen characterization and sperm storage in *Cabot's Tragopan*. Poult Sci 2006; 85(5): 892-8. [http://dx.doi.org/10.1093/ps/85.5.892] [PMID: 16673768]

[8] Adams EM, Wolfner MF. Seminal proteins but not sperm induce morphological changes in the *Drosophila melanogaster* female reproductive tract during sperm storage. J Insect Physiol 2007; 53(4): 319-31. [http://dx.doi.org/10.1016/j.jinsphys.2006.12.003] [PMID: 17276455]

[9] Holt WV, Lloyd RE. Sperm storage in the vertebrate female reproductive tract: how does it work so well? Theriogenology 2010; 73(6): 713-22. [http://dx.doi.org/10.1016/j.theriogenology.2009.07.002] [PMID: 19632711]

[10] Roy VK, Krishna A. Sperm storage in the female reproductive tract of Scotophilus heathii: role of androgen. Mol Reprod Dev 2011; 78(7): 477-87. [http://dx.doi.org/10.1002/mrd.21328] [PMID: 21638510]

[11] Racey PA. The prolonged survival of spermatozoa in bats. In: Duckett JG, Racey PA, Eds. The biology of the male gamete. London: Academic Press 1975; pp. 385-416.

[12] Racey PA. The prolonged storage and survival of spermatozoa in Chiroptera. J Reprod Fertil 1979; 56(1): 391-402.
[http://dx.doi.org/10.1530/jrf.0.0560391] [PMID: 381655]

[13] Racey PA. Ecology of bat reproduction. In: Kunz TH, Ed. Ecology of Bats). New York: Plenum Press 1982; pp. 57-104.
[http://dx.doi.org/10.1007/978-1-4613-3421-7_2]

[14] Wang Z, Liang B, Racey PA, Wang Y, Zhang S. Sperm storage, delayed ovulation, and menstruation of the female Rickett, s big- footed Bat (*Myotis ricketti*). Zool Stud 2008; 47: 215-21.

[15] Crichton EG, Krutzsch PH. Reproductive biology of the female little mastiff bat, *Mormopterus planiceps* (Chiroptera: Molossidae) in southeast Australia. Am J Anat 1987; 178(4): 369-86.
[http://dx.doi.org/10.1002/aja.1001780408] [PMID: 3604956]

[16] Wimsatt WA. Some interrelations of reproduction and hibernation in mammals. Symp Soc Exp Biol 1969; 23: 511-49.
[PMID: 5367180]

[17] Gopalakrishna A, Madhavan A. Survival of spermatozoa in the female genital tract of the Indian vespertilionid bat, *Pipistrellus ceylonicus chrysothrix* (Wroughton). Proc Indiana Acad Sci 1971; 73: 43-9.

[18] Racey PA. The viability of spermatozoa after prolonged storage by male and femal European bats. Period Biol 1973; 75: 201-5.

[19] Doak RL, Hall A, Dale HE. Longevity of spermatozoa in the reproductive tract of the bitch. J Reprod Fertil 1967; 13(1): 51-8.
[http://dx.doi.org/10.1530/jrf.0.0130051] [PMID: 6066781]

[20] Martinet L, Raynaud F. Survie prolongée des spermatozoïdes dans l'utérus de la Hase: explication de la superfoetation. In: Hafez ES, Thibau CG, Eds. Sperm Transport, Survival and Fertilising Ability. Paris: INSERM 1974; pp. 295-308.

[21] Krishna A, Dominic CJ. Storage of spermatozoa in the female genital tract of the vespertilionid bat, Scotophilus heathi. J Reprod Fertil 1978; 54(2): 319-21.
[http://dx.doi.org/10.1530/jrf.0.0540319] [PMID: 569203]

[22] Krishna A. Storage of spermatozoa in the female genital tract of the Indian pigmy pipistrelle bat, *Pipistrellus mimus Wroughton.* Archs Biol Bruxelles 1984; 95: 223-9.

[23] Holt WV, Fazeli A. Sperm storage in the female reproductive tract. Annu Rev Anim Biosci 2016; 4: 291-310.
[http://dx.doi.org/10.1146/annurev-animal-021815-111350] [PMID: 26526545]

[24] Taggart DA, Temple-Smith PD. Transport and storage of spermatozoa in the female reproductive tract of the brown marsupial mouse, *Antechinus stuartii* (Dasyuridae). J Reprod Fertil 1991; 93(1): 97-110.
[http://dx.doi.org/10.1530/jrf.0.0930097] [PMID: 1920303]

[25] Uchida TA, Mori T. Prolonged storage of spermatozoa in hibernating bats. In: Fenton MB, Racey PA, Rayner JM, Eds. Recent Advances in the Study of Bats. Cambridge: Cambridge University Press 1987; pp. 351-65.

[26] Kitchener DJ. Reproduction in female Gould's Wattled Bat Chalinolobus gouldii (Gray) (Vespertilionidae) in Western Australia. Aus J Zool 1975; 23: 29-42.

[27] Mori T, Uchida TA. Sperm storage in the reproductive tract of the female Japanese long-fingered bat, *Miniopterus schreibersii fuliginosus.* J Reprod Fertil 1980; 58(2): 429-33.
[http://dx.doi.org/10.1530/jrf.0.0580429] [PMID: 7191895]

[28] Mori T, Oh YK, Uchida TA. Sperm storage in the oviduct of the Japanese greater horseshoe bat, *Rhinolophus ferrumequinu nippon.* J Fac Agric Kyushu Univ 1982; 27: 47-53.

[29] Smith TT, Yanagimachi R. The viability of hamster spermatozoa stored in the isthmus of the oviduct: the importance of sperm-epithelium contact for sperm survival. Biol Reprod 1990; 42(3): 450-7. [http://dx.doi.org/10.1095/biolreprod42.3.450] [PMID: 2340331]

[30] Krishna A. The relationship between spermatozoa and epithelium of the female genital tract during sperm storage in the greater yellow bats (*Scotophilus heathi*): the light and electronmicroscopic observations. Proc Natl Sci Counc Repub China B 1997; 21(1): 31-6. [PMID: 9208484]

[31] Suarez SS, Katz DF, Owen DH, Andrew JB, Powell RL. Evidence for the function of hyperactivated motility in sperm. Biol Reprod 1991; 44(2): 375-81. [http://dx.doi.org/10.1095/biolreprod44.2.375] [PMID: 2009336]

[32] Tingari MD, Lake PE. Ultrastructural studies on the uterovaginal sperm-host glands of the domestic hen, *Gallus domesticus.* J Reprod Fertil 1973; 34(3): 423-31. [http://dx.doi.org/10.1530/jrf.0.0340423] [PMID: 4741313]

[33] Holt WV. Does apoptosis hold the key to long-term sperm storage mechanisms *in vivo?* Mol Reprod Dev 2011; 78(7): 464-5. [http://dx.doi.org/10.1002/mrd.21347] [PMID: 21721067]

[34] Fawcett DW, Ito S. The fine structure of bat spermatozoa. Am J Anat 1965; 116: 567-609. [http://dx.doi.org/10.1002/aja.1001160306] [PMID: 14324688]

[35] Uchida TA, Mori T. Electron microscope studies on the fine structure of germ cells in Chiroptera. Spermiogenesis in some bats and notes on its phylogenetic significance. Sci Bull Fac Agrie Kyushu Univ 1972; 26: 399-418.

[36] Krutzsch PH, Crichton EG, Nagle RB. Studies on prolonged spermatozoa survival in Chiroptera: a morphological examination of storage and clearance of intrauterine and cauda epididymal spermatozoa in the bats *Myotis lucifugus* and *M. velifer*. Am J Anat 1982; 165(4): 421-34. [http://dx.doi.org/10.1002/aja.1001650406] [PMID: 7158612]

[37] Uchida TA, Mori T, Oh YK. Sperm invasion of the oviducal mucosa, fibroblastic phagocytosis and endometrial sloughing in the Japanese greater horseshoe bat, Rhinolophus ferrumequinum nippon. Cell Tissue Res 1984; 236(2): 327-31. [http://dx.doi.org/10.1007/BF00214234] [PMID: 6733759]

[38] Crichton EG, Krutzsch PH, Wimsatt WA. Studies on prolonged spermatozoa survival in Chiroptera. I. The role of uterine free fructose in the spermatozoa storage phenomenon. Comp Biochem Physiol 1981; 70A: 387-95. [http://dx.doi.org/10.1016/0300-9629(81)90195-X]

[39] Crichton EG, Krutzsch PH, Chavpil M. Studies on prolonged spermatozoa survival in Chiroptera. II. The role of zinc in the spermatozoa storage phenomenon. Comp Biochem Physiol 1982; 71-7. [http://dx.doi.org/10.1016/0300-9629(82)90369-3]

[40] Crichton EG, Krutzsch PH, Yanagimachi R. Stability of the sperm plasma membrane of hibernating bats (*Myotis velifer*) compared with other mammals. J Reprod Fertil 1993; 97(1): 1-4. [http://dx.doi.org/10.1530/jrf.0.0970001] [PMID: 8463999]

[41] Krutzsch P, Crichton EG, Lennon DL, Stratman FW, Carter AL. Studies on prolonged spermatozoa survival in chiroptera. III. Preliminary data on carnitine. Andrologia 1984; 16(1): 34-7. [http://dx.doi.org/10.1111/j.1439-0272.1984.tb00231.x] [PMID: 6721238]

[42] Lambert H. Temperature dependence on capacitation in bat sperm monitored by zona- free bat ova. Gamete Res 1981; 4: 525-33. [http://dx.doi.org/10.1002/mrd.1120040606]

[43] Uchida TA, Mori T Prolonged storage of spermatozoa in hibernating bats. In: Fenton MB, Racey PA, Rayner JMV, Eds. Recent Advances in the Study of Bats. Cambridge: Cambridge University Press 1987; pp. 351-6.

[44] Hartman CG. Ovulation, fertilization and the transport and viability of eggs and spermatozoa. In: Allen E, Danforth CH, Doisy EA, Eds. Sex and Internal Secretions. Baltimore: Williams & Wilkins 1939; pp. 630-719.

[45] Pagenstecher HA. Über die Begattung von Vesperugo pipistrellus. Verh. naturh.-med. Ver Heidelb 1859; 1: 194-5.

[46] Uchida TA. Fertilization and hibernation in bats. Heredity, Tokyo 1957; 11: 14-7.

[47] Wimsatt WA. Some problems of reproduction in relation to hibernation in bats. Bull Mus comp Zool Harv 1960; 124: 249-70.

[48] Hartman CG. On the survival of spermatozoa in the female genital tract of the bat. Q Rev Biol 1933; 8: 185-93.
[http://dx.doi.org/10.1086/394433]

[49] Medway L. Reproductive cycles of the flat-headed bats *Tylonycteris pachypus* and *T. robustula* (Chiroptera: Vespertilionidae) in a humid equatorial environment. Zool J Linn Soc 1972; 51: 33-61.
[http://dx.doi.org/10.1111/j.1096-3642.1972.tb00772.x]

[50] Rasweiler JJ IV, Badwaik NK, Mechineni KV. Selectivity in the transport of spermatozoa to oviductal reservoirs in the menstruating fruit bat, Carollia perspicillata. Reproduction 2010; 140(5): 743-57.
[http://dx.doi.org/10.1530/REP-10-0130] [PMID: 20696851]

[51] Hoffman LH, Wimsatt WA. Histochemical and electron microscopic observations on the sperm receptacles in the garter snake oviduct. Am J Anat 1972; 134(1): 71-95.
[http://dx.doi.org/10.1002/aja.1001340107] [PMID: 4260659]

[52] Van Krey HP, Ogasawara FX, Pangborn J. Light and electron microscopic studies of possible sperm gland emptying mechanisms. Poult Sci 1967; 46(1): 69-78.
[http://dx.doi.org/10.3382/ps.0460069] [PMID: 6067789]

[53] Andreuccetti P, Angelini F, Taddei C. The interactions between spermatozoa and uterine epithelium in the hibernating bat, *Pipistrellus kuhli* natt. Gamete Res 1984; 10: 67-76.
[http://dx.doi.org/10.1002/mrd.1120100108]

[54] Nakano O. Über die Verteilung des Glykogens bei den zyklischen Veränderungen in den Geschlechtsorganen der Fledermaus, und über die Nahrungsaufnahme der Spermien in den weiblichen Geschlechtswege. Folia Anat Jap 1928; 6: 777-828.

[55] Hunter AG, Barker LD, Johnson WL, Fahning ML, Schultz RH. Antigenicity, toxicity and cross-reactions of male bat (*Myotis lucifugus*) reproductive organs. J Reprod Fertil 1971; 24(2): 171-7.
[http://dx.doi.org/10.1530/jrf.0.0240171] [PMID: 4994571]

[56] Hunter AG, Johnson WL, Barker LD, Fahning ML, Schultz RH. Bat seminal vesicle protein: its characterization and physiological properties. J Reprod Fertil 1971; 24(2): 179-86.
[http://dx.doi.org/10.1530/jrf.0.0240179] [PMID: 5102534]

[57] Crichton EG. Sperm storage and fertilization. In: Crichton EG, and P.H. Krutzsch PH, Eds. Reproductive biology of bats. London, Academic Press 2000; 295-320.
[http://dx.doi.org/10.1016/B978-012195670-7/50008-4]

[58] Roy VK, Krishna A. Evidence of androgen-dependent sperm storage in female reproductive tract of *Scotophilus heathi*. Gen Comp Endocrinol 2010; 165(1): 120-6.
[http://dx.doi.org/10.1016/j.ygcen.2009.06.012] [PMID: 19539620]

[59] Jones R. Sperm survival *versus* degradation in the Mammalian epididymis: a hypothesis. Biol Reprod 2004; 71(5): 1405-11.
[http://dx.doi.org/10.1095/biolreprod.104.031252] [PMID: 15215193]

[60] Roy VK, Krishna A. Changes in glucose and carnitine levels and their transporters in utero-tubal junction in relation to sperm storage in the vespertilionid bat, *Scotophilus heathi.* J Exp Zool A Ecol Genet Physiol 2013; 319(9): 517-26.
[PMID: 24039231]

[61] Storey BT. Mammalian sperm metabolism: oxygen and sugar, friend and foe. Int J Dev Biol 2008; 52(5-6): 427-37.
[http://dx.doi.org/10.1387/ijdb.072522bs] [PMID: 18649255]

[62] Ford WC. Glycolysis and sperm motility: does a spoonful of sugar help the flagellum go round? Hum Reprod Update 2006; 12(3): 269-74.
[http://dx.doi.org/10.1093/humupd/dmi053] [PMID: 16407453]

[63] Krutzsch PH, Crichton EG. Fertilization in Bats. A Comparative Overview of Mammalian Fertilization. New York: Plenum Press 1991.
[http://dx.doi.org/10.1007/978-1-4757-8982-9_8]

[64] Watson PF. Recent developments and concepts in the cryopreservation of spermatozoa and the assessment of their post-thawing function. Reprod Fertil Dev 1995; 7(4): 871-91.
[http://dx.doi.org/10.1071/RD9950871] [PMID: 8711221]

[65] Eisenbach M. Why are sperm cells phagocytosed by leukocytes in the female genital tract? Med Hypotheses 2003; 60(4): 590-2.
[http://dx.doi.org/10.1016/S0306-9877(03)00054-9] [PMID: 12615529]

[66] Coy P, Garcia-Vazquez FA, Visconti PE, Aviles M. Roles of the oviduct in mammalian fertilization. Reproduction 2012; 144(6): 649-60.

[67] Austin CR. Observations on the penetration of the sperm in the mammalian egg. Aust J Sci Res, B 1951; 4(4): 581-96.
[http://dx.doi.org/10.1071/BI9510581] [PMID: 14895481]

[68] Chang MC. Fertilizing capacity of spermatozoa deposited into the fallopian tubes. Nature 1951; 168(4277): 697-8.
[http://dx.doi.org/10.1038/168697b0] [PMID: 14882325]

[69] Moein-Vaziri N, Phillips I, Smith S, *et al.* Heat-shock protein A8 restores sperm membrane integrity by increasing plasma membrane fluidity. Reproduction 2014; 147(5): 719-32.
[http://dx.doi.org/10.1530/REP-13-0631] [PMID: 24501193]

[70] Abhilasha KA, Krishna A. Adiposity and androstenedione production in relation to delayed ovulation in the Indian bat, *Scotophilus heathi.* Comp Biochem Physiol C Pharmacol Toxicol Endocrinol 1997; 116(1): 97-101.
[http://dx.doi.org/10.1016/S0742-8413(96)00142-9] [PMID: 9080676]

[71] Abhilasha KA, Krishna A. High androgen production by ovarian thecal interstitial cells: a mechanism for delayed ovulation in a tropical vespertilionid bat, *Scotophilus heathi.* J Reprod Fertil 1996; 106(2): 207-11.
[http://dx.doi.org/10.1530/jrf.0.1060207] [PMID: 8699403]

[72] Le Y, Chen S, Hu L, *et al.* B-cell lymphoma-2 localization in the female reproductive tract of the Chinese soft-shelled turtle, Pelodiscus sinensis and its relationship with sperm storage. Anat Rec (Hoboken) 2015; 298(12): 2011-7.
[http://dx.doi.org/10.1002/ar.23258] [PMID: 26285642]

[73] Liu T, Chu X, Huang Y, *et al.* Androgen-related sperm storage in oviduct of Chinese Soft-Shelled Turtle *in vivo* during annual cycle. Sci Rep 2016; 6: 20456.
[http://dx.doi.org/10.1038/srep20456] [PMID: 26847578]

[74] Ashizawa K, Nishiyama H, Nagae T. Effects of oviducal cells on the survival and fertilizing ability of fowl spermatozoa. J Reprod Fertil 1976; 47(2): 305-11.
[http://dx.doi.org/10.1530/jrf.0.0470305] [PMID: 957330]

[75] Ashizawa K, Nishiyama H. Effects of various cultured cells on the survival and fertilizing ability of fowl spermatozoa. J Reprod Fertil 1977; 49(2): 405-7. [http://dx.doi.org/10.1530/jrf.0.0490405] [PMID: 850230]

[76] Ashizawa K, Tokudome Y, Okauchi K, Nishiyama H. Effects of HeLa and BHK-21 cells on the survival of fowl, bull, ram and boar spermatozoa *in vitro*. J Reprod Fertil 1982; 66(2): 663-6. [http://dx.doi.org/10.1530/jrf.0.0660663] [PMID: 6890992]

[77] Ashizawa K, Nishiyama H. Prolonged survival of fowl spermatozoa in the oviducal tissues in organ culture. Br Poult Sci 1983; 24(1): 27-32. [http://dx.doi.org/10.1080/00071668308416710] [PMID: 6831275]

[78] Ashizawa K, Tamiya E, Okauchi K, Nishiyama H. Prolonged motility of fowl spermatozoa *in vitro* due to a low-molecular weight factor(s) released from cultured embryonic cells. Anim Reprod Sci 1985; 9: 181-8. [http://dx.doi.org/10.1016/0378-4320(85)90022-3]

[79] Sharifi M, Ghiorbani R, Fazeli A, Holt W. Evidence of sperm storage in *Pipistrellus kuhlii* (Chiroptera: Vespertilionidae) in western Iran. Folia zoological 2004; 53: 1-6.

[80] Boilard M, Reyes-Moreno C, Lachance C, *et al.* Localization of the chaperone proteins GRP78 and HSP60 on the luminal surface of bovine oviduct epithelial cells and their association with spermatozoa. Biol Reprod 2004; 71(6): 1879-89. [http://dx.doi.org/10.1095/biolreprod.103.026849] [PMID: 15286042]

Frontiers in Reproductive Science, 2017, *Vol. 1*, 119-147

CHAPTER 7

Sperm Competition and Cryptic Female Choice in Bats

C. Ruth Archer[1], **Teri J. Orr**[2] and **David J. Hosken**[1,*]

[1] *Centre for Ecology and Conservation, College of Life and Environmental Sciences, University of Exeter, Tremough Campus, TR10 9EZ, Penryn, UK*

[2] *Department of Biology, University of Utah, Salt Lake City, UT, USA*

Abstract: Some of the most striking behaviours and characters in nature are the result of sexual selection, including sexual dimorphism in size, pigmentation, ornamentation and weaponry. Our understanding of sexual selection is largely based on birds and insects research and, of the mammals, primates and ungulates. Bats have received comparatively little attention. Indeed, bats may not appear to be the most obvious model for studying sexual selection because males lack the exaggerated weaponry seen in some mammals, and typically lack visually striking ornamental traits. However, bats are well placed for studying two key aspects of sexual selection, sperm competition and cryptic female choice. Here we introduce these processes, before explaining why they may be particularly important in bats. We then review research on sperm competition and cryptic female choice in bats and highlight the tremendous opportunities to improve our understanding further by studying the Chiroptera.

Keywords: Bats, Chiroptera, Cryptic female choice, Sexual conflict, Sexual selection, Sperm competition.

SEXUAL SELECTION

In many animals, the sexes look so different that they could be mistaken for separate species [1, 2]. Males often have bright pigmentation, elaborate ornaments and behaviours, while females are inconspicuous and lack ornamentation. These striking male characters are clearly costly, as developing and expressing them is energetically demanding and the very nature of these characters make males conspicuous, elevating predation risk. These obvious naturally selected costs together with sexual dimorphism in trait expression led Darwin [3] to develop his theory of sexual selection. He suggested that although these highly conspicuous characters are not favoured by natural selection, they are important in reproduc-

[*] **Corresponding author David J. Hosken:** Centre for Ecology and Conservation, College of Life and Environmental Sciences, University of Exeter, Tremough Campus, TR10 9EZ, UK; Tel: +44 (0)1326 371834; E-mail: D.J.Hosken@exeter.ac.uk

Edith Arenas-Ríos (Ed.)

tive competition (sexual selection). Therefore, while natural selection favours traits that improve overall survival, sexual selection promotes traits that increase mating success.

Sexual selection is therefore defined as the reproductive competition that occurs between members of the same sex and same species for access to mates (or more strictly, fertilizations) [3]. It requires variance in mating success, although variance in mating success need not equal selection, and typically, the variance in mating success is greater in males than females. The reason for this is rooted in anisogamy [4]: females produce fewer, larger gametes (ova) than males, who produce many, tiny gametes (sperm). This inequality in gamete production means that a female's reproductive success is limited by how many eggs she can produce, while males are constrained by how many females they can successfully inseminate with their large sperm stores [5]. This means that to improve their reproductive success, males typically compete to mate with as many females as possible, while females choose from these competitors to mate with a subset of high quality or highly attractive males [1, 6]. These two mechanisms, male-male competition and female mate-choice, typically drive sexual selection. While we focus on this typical scenario where males compete more for mates and females are the choosier sex, it is worth noting that in some species, sex-role reversal occurs [*e.g.*, 7].

Sexual selection is now widely recognised as a major cause of evolution and male-male competition and female mate choice are jointly accepted as mechanisms underpinning this form of selection. However, it has become clear that rather than females simply picking mates and males competing for mating, both competition and choice can extend after mating in the form of sperm competition and cryptic female choice.

POST-COPULATORY SEXUAL SELECTION: COMPETING SPERM AND CRYPTIC CHOICE

Even after a male has mated, fertilization success is not guaranteed. This is because females often mate with multiple males (*i.e.* polyandry) [8, 9] meaning that in species with internal fertilization, sperm from different males can co-occur in a female's reproductive tract. In an extreme example, Honeybee queens (*Apis mellifera*) mate on a single nuptial flight with several males, whose sperm are then stored for up to 7 years [10]. This creates a second opportunity for male-male competition as the sperm from different males compete to fertilize ova [sperm competition - 11]. Females can also bias the outcome of this competition in favour of a particular male by cryptically selecting which males enjoy the greatest fertilization success (cyptic female choice - [12 , 13]).

If we assume for now that females simply provide an arena for sperm competition to take place and sperm competition proceeds *via* the raffle principle, then each competing sperm has an equal chance of fertilization. In this case, paternity share reflects the proportion of sperm in the fertilization set [14]. This appears to be the case in leaf cutter ants (*Atta colombica*), where fertilization is determined by how many sperm males transfer at mating [15]. When sperm competition success is determined by a male's numerical representation in the group of sperm competing for fertilization, selection should favour the production of large ejaculates [14]. Males may even transfer so much sperm at mating that they displace the sperm of their rivals from where they are stored within females [16, 17].

Fertilization success does not always reflect a fair raffle however, and often particular sperm/ejaculates have an advantage over rivals [18]. In other words, the sperm competition raffle can be loaded to favour certain ejaculates [14]. This could be because the probability of fertilization is weighted by mating order, as in the sierra dome spider (*Linyphia litigiosa*), where the first male to mate has a fertilization advantage over subsequent males [19], while the reverse is true in Scorpionflies, (*Harpobittacus nigriceps*), where the last male has improved fertilization odds [20]. In any case, because ejaculates are costly to produce, selection should favour males that can strategically adjust their ejaculate transfer to improve their fertilization success [21].

The outcome of sperm competition may not just depend on ejaculate size and the timing of insemination, but on some quality of the sperm itself and numerous sperm traits, including viability, lifespan and size, can all affect fertilization success [18]. This may simply be because a particular trait gives sperm a competitive advantage or instead, because these traits are selected for by cryptic female choice (reviewed in - [13]).

Here we define cryptic female choice as non-random variance in fertilization success after mating, due to females [22]. There are multiple mechanisms by which females can exert cryptic choice [12]. In chinook salmon for example, females can vary the composition of their ovarian fluid to differentially improve the swimming speed of sperm from particular mates [23]. While in feral fowl, females actively dump non-preferred sperm out of their reproductive tracts [24]. Reproductive tract morphology can also be important as is the case in the fly *Drosophila simulans*. Here females have two types of sperm-storage organs, the spermathecae and the seminal receptacle. The seminal receptacle preferentially stores the sperm from first mating males, while the spermathecae favours sperm from second males, and females are able to utilise sperm from one or other organ depending on which is their preferred mate [25, 26].

Collectively, cryptic female choice and sperm competition explain why fertilization success can be biased towards the sperm of genetically compatible [27], diverse [9], unrelated [28, 29], viable or attractive [30] males. These processes are therefore clearly important aspects of sexual selection [11, 31, 32]. One mammalian group in which post-copulatory sexual selection is likely to be very important are the bats, because females frequently store sperm for prolonged periods [33], may have little opportunity for pre-copulatory mate choice [34] and monogamy is uncommon [35].

BAT REPRODUCTION

As noted above, bats have many traits that make them likely to experience significant post-copulatory sexual selection. Firstly, bats have diverse mating systems, with species that tend towards monogamy (single male, single female mating groups), polygyny (single male, multiple female mating groups) and promiscuity (multiple male, multiple female groups) [35]. Several lines of evidence suggest that polyandry is likely to widespread in bats [35, 36]. For example, in many temperate and tropical species male mating territories are seasonal and females often move between different mating groups, giving females the opportunity to mate with multiple males [35]. In several species of hibernating bats, males mate with torpid females [37, 38], which may also lead to females mating with multiple males, and sperm from different males have been detected in the reproductive tracts of female bats [36]. Similarly, although singleton young are the norm for bats [39] examination of several taxa that bear twins has revealed multiple paternity [40, 41]. So in many bats, females are clearly mating with several males.

Polyandry only provides the opportunity for sperm competition if the sperm of different males co-occurs in the female's reproductive tract. In many bats this is inevitable. Bats are one of the very few mammalian groups where females store sperm for prolonged periods after mating [33, 42 - 45] (Fig. **1**). For example, many temperate species store sperm transferred during mating in autumn, until spring when fertilization occurs [44, 46, 47]. Sperm storage clearly sets the stage for sperm competition and for cryptic female choice. However, bats are notable among mammals as the only Order exhibiting all three forms of reproductive delay: delayed fertilization, implantation and development [48] (Table **1**). Delaying implantation and development allows further opportunities for females to exert cryptic choice [49]. In the case of *Myotis lucifugus* and many other temperate species males also store sperm for extended periods of time [50]. Since males often mate months after spermatogenesis is completed [51] there may well be very strong selection on sperm longevity [136] and other characters, including strategic ejaculation.

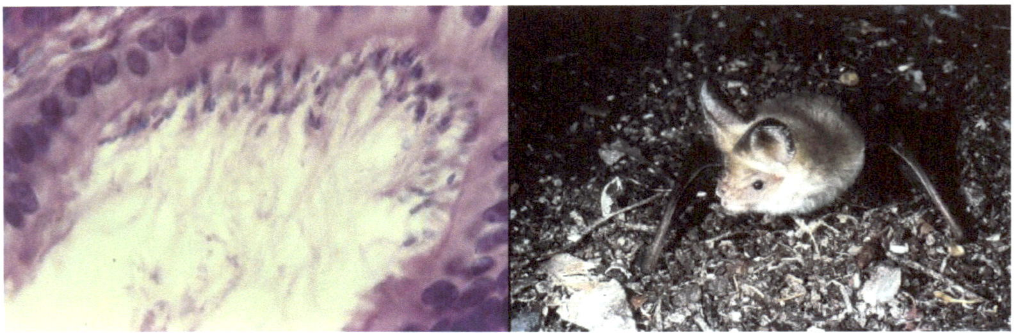

Fig. (1). The sperm-storing Australian lesser long-eared bat *Nyctophilus geoffroyi*. The left-panel shows a histological section approximately though the utero-tubual junction, stained with eosin and haematoxylin, in which stored sperm are clearly visible with heads touching the epithelia cells. The right panel shows a living lesser long-eared bat.

Table 1. Delineated opportunities for pre and post-copulatory sexual selection facilitated by the different reproductive modes described in bats. Reproductive modes include; the lack of delays (no delay), delayed fertilization (sperm storage), delayed implantation, delayed development and the speculated occurrence of superfoetation (overlapping pregnancies and reproductive stages within a single female at a single time; see Roellig *et al.* [109] for a detailed discussion of the difficulties of assigning superfoetation). Types of sexual selection are precopulatory (preC) and post copulatory which is further broken down by general categories; CFC-cryptic female choice, SC-sperm competition, GC-genomic conflict (conflict realised between maternal, and paternal or embryo genes during pregnancy), and SAC-sexually antagonistic co-evolution. While SC and CFC are both likely in taxa with any of the reproductive modes illustrated they may be exaggerated in the case of delays. To illustrate these differences predicted likelihood of occurrence is indicated by predictions; low, medium, high and variable. All the predictions outlined would vary in the context of additional aspects of an animal's biology such as mating system and population density.

Reproductive Mode	Type of Sexual Selection	Likelihood	Specifics
No Delay	PreC	medium/high	limited periods of female receptivity (*e.g.* oestrus, induced ovulation), mating system dependant
	CFC (♀)	low/variable	limited to duration of each reproductive stage
	SC (♂)	low/variable	limited to period of sperm longevity, mating system dependant
	GC (♀♂)	low/variable	limited to duration of relevant reproductive stage
	SAC (♀♂)	low/variable	possible manipulation of female's current reproductive bout by males to reinstate oestrus

(Table 1) contd.....

Reproductive Mode	Type of Sexual Selection	Likelihood	Specifics
Delayed Fertilization (Sperm Storage)	PreC	medium/high	increased period of receptivity to mating opportunities, mating system dependant
	CFC (♀)	high	increased opportunity for sperm dumping, copulatory plug removal, selective sperm storage and/or phagocytosis
	SC (♂)	high	increased opportunity for manipulation of females *via* the ejaculate, including copulatory plug formation
	GC (♀♂)	low/medium	limited GC based on contents of the ejaculate and/or sperm phenotype
	SAC (♀♂)	high	increased opportunity for conflict over sperm storage and use
Delayed Implantation	PreC	medium	enhanced period of receptivity to mating opportunities, mating system dependant
	CFC (♀)	high	increased opportunity for selective reabsorption or implantation of blastocysts
	SC (♂)	medium	possible SC depending on mating system
	GC (♀♂)	high	increased GC between maternal and paternal genes expressed by blastocysts
	SAC (♀♂)	high	limited implantation sites or female control over implantation may lead to SAC between maternal and paternal genes
Delayed Development	PreC	medium	limited receptivity to periods of oestrus and/or induced ovulation, mating system dependant
	CFC (♀)	high	increased opportunity for female evaluation of embryo genomes, selective embryo reabsorption and/or provisioning
	SC (♂)	medium	depends on mating system
	GC (♀♂)	high	increased GC between maternal and paternal genes expressed by developing embryo
	SAC (♀♂)	high	exaggerated by increased duration of development and female control of embryo provisioning during delay
Superfoetation	PreC	low to high	enhanced receptivity to mating opportunities, mating system dependant
	CFC (♀)	high	increased opportunity to select multiple sires as mates become available independent of ongoing reproductive stages

(Table 1) contd.....

Reproductive Mode	Type of Sexual Selection	Likelihood	Specifics
Superfoetation	SC (♂)	high	increased SC if superfoetation presents a longer period of sperm maintenance by females, mating system dependant
	GC (♀♂)	high	enhanced due to lengthened and co-occurring reproductive events which may even lead to conflict between ♂s who have offspring at the same or different stages
	SAC (♀♂)	high	increased opportunities for numerous mechanisms of SAC to occur over co-occurring stages of reproduction

Post-copulatory sexual selection may be particularly important in bats because in some species, females appear to lack pre-copulatory mate choice. For example, in little brown bats (*M. lucifugus*) mating often occurs when females are in torpor [37]. Similarly, in the Brazilian free-tailed bat (*Tadarida brasiliensis*) males can force copulation or mate while females sleep, suggesting little or no mate assessment by females [52]. In these cases, females may only be able to choose sires after mating. However, the frequency of these 'forced matings', and if such matings result in paternity, is unknown.

Clearly there is enormous scope for sperm competition and cryptic female choice in bats and a picture is beginning to emerge of the importance and evolutionary implications of post-copulatory sexual selection in this group. Here we review our current understanding. We first focus on how post-copulatory sexual selection has shaped investment in sperm production (testis size), sperm morphology, the composition of fluid delivered alongside sperm and genital morphology in both sexes. We then discuss what we know about cryptic female choice in this taxon, and finally, what is left to discover about sperm competition and cryptic female choice in bats. We hope to highlight the tremendous research opportunities that exist for improving our understanding of sperm competition and cryptic female choice in this the second most species-rich mammalian Order.

THE EVOLUTIONARY CONSEQUENCES OF SPERM COMPETITION IN BATS

Sperm Competition and Sperm Numbers

A clear prediction of sperm competition theory is that if sperm competition follows the raffle principle (as it appears to in many mammals), then males should make more sperm as the probability of sperm competition (sperm competition risk) increases [53]. This greater investment in sperm production typically results in the evolution of larger testes. This has been shown experimentally in yellow dung flies (*Scathophaga stercoraria*) where flies evolved larger testes under increased sperm competition risk [54]. More generally, comparative studies find a positive association between testis size and sperm competition risk in many taxa [13, 55, 56].

In bats, testes account for between 0.12% to 8.4% of male body mass and so, relative to their body size, bats have the biggest testes of any mammalian order and also exhibit the broadest range of relative testis sizes [36]. Several studies have related this variation to different indicators of sperm competition risk and found that testes are larger in species with more intense sperm competition (Fig. **2**). Across all bats (both pteropodidae [57] and non-pteropodidae [38]), males that live in larger groups (where females have more opportunities to mate multiply) have relatively larger testes. Testis size also varies between species with different mating systems as might be predicted by sperm competition theory. Testes are largest in species where multiple males roost with multiple females, intermediate in species were a single male mates with multiple females and smallest in monogamous species [36]. In species with delayed fertilization, where there is a high chance that sperm from different males overlap in the female reproductive tract, males also have larger testes [58].

These data show very clearly that in bats testis size positively correlates with three different indicators of sperm competition risk; group size, mating system, and sperm storage (delayed fertilization). This provides clear support for sperm competition selecting on investment in sperm production, resulting in increased testis mass. It also means that testis size is a reliable indicator of sperm competition risk in bats.

Sperm Size, Structure and Swimming Speed

Fertilization success could be influenced by sperm morphometry as well as sperm numbers. Mammalian sperm are made up of a head, mid-piece and a flagellum. The flagellum, or tail, is at least partly responsible for sperm motility within the female reproductive tract, while the mid-piece contains mitochondria that fuel the sperm's movement (although it is often unclear if glycolysis also contributes to ATP production, either in part or solely). Finally, the head contains the sperm's nucleus and is surrounded by an acrosome, which is packed with the enzymes that are heavily implicated in gamete fusion. The shape and dimensions of each of these units varies across and within taxa and this variation may be due in part, to the influence of sperm competition (reviewed in, [59]). This may be particularly true in temperate bats because in many of these species, mating occurs long after the testes cease making sperm [50]. This means that males transfer aged sperm at mating because males cannot replace older and potentially suboptimal gametes though continuous spermatogenesis (Fig. **2**). This may increase selection on "sperm quality" rather than quantity in these temperate species. The problem is however, that it is not always clear what sperm characteristics determine quality.

Sperm competition may simply favour larger male gametes, which fill female sperm storage organs (or in the case of bats storage regions [60]) and prevent females from storing the sperm of rival males [61, 62]. However, while species with greater sperm competition risk often have larger or longer sperm, in other species sperm size is independent of the risk of sperm competition (Table **2**).

Table 2. Association between sperm morphology and the risk of sperm competition across a range of taxa. Symbols: + = positive relationship, / = no relationship and - = negative relationship. This table is not exhaustive but is intended to illustrate the association between sperm morphology and the risk of sperm competition varies between different classes and between families belonging to the same class. Taxonomic descriptions are the same as those used in the reference cited. n = number of species in comparison.

Taxon	n	Sperm Trait	Relationship to Sperm Competition	Ref.
Amphibians				
Frogs	100	Head length Tail length	+ +	[110]
Birds				
Passerines	22	Variation in sperm length	-	[111]
Passerines	42	Total length	+	[62]
Passerines	55	Total length Variation in sperm length	+ -	[112]
New World Blackbirds	21	Total length	+	[113]
Finches	18	Head length Mid-piece length Flagellum length Total length	/ + + +	[114]
Old World Warblers	22	Head length Mid-piece length Flagellum length Total length	/ - - -	[114]
Fish				
26 families	89	Total length	-	[56]
Cichlids	21	Total length	+	[115]
Cichlids	29	Total length	+	[66]
Insects				
Callosobruchus maculatus	1	Total length	/	[116]
Butterflies	74	Total length	+	[55]

(Table 2) contd.....

Taxon	n	Sperm Trait	Relationship to Sperm Competition	Ref.
Moths	178	Total length	+	[117]
Eusocial bees & ants	27	Sperm length Variation in length	/ -	[118]
Mammals				
Eutherians	226	Total sperm length Head length Mid-piece length Principal piece length Total flagellum length Head length / Head width	+ + / + + +	[119]
Marsupials	28	Sperm length Head length Mid-piece length Principal piece length Total flagellum length Head length / Head width	+ / / + + /	[120]
71 genera	123	Head length Head volume Mid-piece length Mid-piece volume Flagellum length Flagellum volume	/ / / + / /	[121]
Reptiles				
Snakes	25	Total length Head length Mid-piece length Principal-piece length Area of fibrous sheath Area of ODF 3 and 8	+ + + + / +	[122]

Sperm competition may instead select for faster sperm [62 - 66]. Initially, researchers assumed that larger sperm, with elongated flagella swim faster. But this is not always the case: in mammals sperm with long flagella swim particularly fast [67, 68] but the same trend is not evident in some bird species [62]. Swimming speed appears to depend on the balance between the force generated by the flagella and the drag imposed by the sperm head and mid-piece [69]. This may mean that sperm swimming speed depends on the ratio of the flagellum to the head [69]. In support of this, sperm competition appears to favour streamlined sperm with elongated heads and / or condensed mid-pieces in red deer (*Cervus elaphus hispanicus*) [68]. However, selection on sperm size and structure will also depend on female effects and the shape of the female reproductive tracts [69]. Consistent with this idea, in mammals sperm shape varies as a function of female reproductive tract morphology: where females have longer or more convoluted oviducts and sperm have to swim further and longer to reach ova, sperm tend to have larger mid-pieces [70]. If mid-piece size reflects mitochondrial number, by having larger mid-sections, sperm may be better provisioned and so able to survive the long swim towards ova [70]. Variation in female reproductive tract morphology may mean that no single sperm shape is suited to fast swimming across different species [69].

Overall, the relationship between sperm morphology and fertilization success varies enormously across species (Table **2**). This makes it hard to predict how sperm shape or structure might respond to sperm competition in any species, and explaining sperm size variation is an enduring problem for sperm biologists [59]. However across bats, sperm morphology does not vary predictably with the risk of sperm competition. Sperm length is unrelated to group size in all bats [38]. These results are supported by comparative analyses of 83 mammalian species, including bats, which did not find an association between testis mass and overall sperm length [71]. Moreover, testis size was independent of other important sperm traits, including the lengths of the sperm head, mid-piece or flagellum, or mid–piece or mitochondrial volumes [71]. Sperm length is also unrelated to the dimensions of the female reproductive tract in a small sample of Megachiroptera [57]. Sperm competition does not appear to select on swimming speed in the small number of bat species sampled either. In Seba's short-tailed fruit bat (*Carollia perspicillata*) where both harem and bachelor males compete for females, males produce slower sperm relative to their testis mass than expected on the basis of interspecific regression [72]. While these data all argue against sexual selection on sperm size or shape, some data suggest that sperm from bat species with delayed fertilization are longer than sperm from species without delayed fertilization [49]. Formal phylogenetically informed analyses of these data are needed to see if delayed fertilization is associated with sperm size variation.

Clearly, information on the effects of sperm competition on sperm morphology in bats is limited, being based on a relatively small number of species and studies, and on a few sperm metrics (*e.g.* sperm length). But overall it appears that while sperm competition has pronounced effects on how many sperm male bats produce, effect sizes for influences on sperm morphometry are smaller. This is probably expected given that sperm numbers seem to play a key role in determining paternity in mammals (*e.g.* [73]).

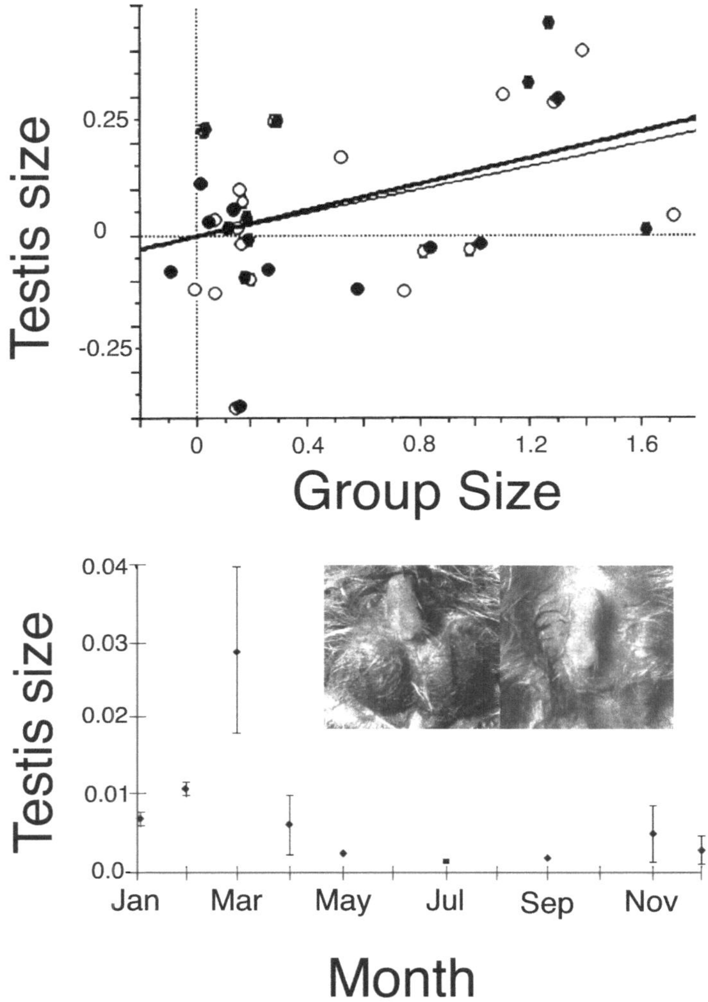

Fig. (2). Bat testis size variation across and within species. Top panel is a plot of testis size variation across the microchiroptera consistent with prediction of sperm competition theory (shown are phylogenetic contrasts in testis size and group size using two different phylogenies [38]. The bottom panel shows seasonal variation in the testis size of an Australian long-eared bat with images of maximally enlarged testes and complete lack of testis development. Note that in this species mating does not seem to occur until about May, well after testes have begun to shrink back to a non-spermatogenic state [51].

Seminal Fluid and Copulatory Plugs

Males do not just transfer sperm at mating, they also transfer seminal fluid. This fluid contains a range of chemicals that have a vast array of physiological effects on sperm and females [21]. For example, seminal fluid may provide sperm with an energy source, buffer sperm from female immune attack [74] or improve sperm swimming speed or orientation [75]. This fluid can even alter female behaviour, elevating female egg laying rates, improving the reproductive success of her current mate for example [76, 77]. Seminal fluid can also reduce how likely females are to remate [78]. Clearly, these attributes mean that seminal fluid is likely to play a key role in post-copulatory sexual selection. In support of this, the glands that make seminal fluid can vary enormously in their size and structure [79] and in some mammals, the size of these glands positively correlates with sperm competition risk [80]. In bats, a large body of evidence suggests that the outcome of sperm competition can be influenced by seminal fluid.

For many bats, fertilization success relies on sperm surviving within the female over winter. These sperm must survive immune attacks [81] and tolerate the acidic pH of the female reproductive tract. This survival can depend on seminal fluid. For example, male *M. lucifugus* transfer a protein in their seminal fluid that is toxic to mice and rabbits, but not to female little brown bats [82]. This protein apparently blocks sperm transport in females and manipulates phagocytes, helping sperm survival [82]. In *M. lucifugus* and *M. velifer* the zinc (Zn) levels in sperm storage sites within females are higher than levels in the same tissues in other mammalian species. This may be partly because females mobilise Zn in these areas, but also because males transfer Zn at mating. This may help improve sperm function and survival given that Zn has important roles in sperm maturation, metabolism and motility [83]. Long-term sperm survival also relies on sperm having a source of energy [84] and seminal fluid may play be vital to provisioning sperm. This appears to be the case in *M. lucifugus* and *M. velifer,* where seminal fluid contributes to the uterine fructose content [85]. Differences in ejaculate composition between taxa with and without long-term sperm storage may provide useful comparisons.

Seminal fluid may also help males succeed in sperm competition by helping create copulatory plugs. In many mammals males leave a rigid plug that fills the vaginal lumen after mating [86]. In some species these plugs prevent female remating [87], but this is not always the case, and in other species mating can occur despite the presence of a plug [88]. In these cases, perhaps the plug helps prevent sperm leakage from the female or make it harder for females to dump sperm [49]. In the case of species with long-term female sperm storage, copulatory plugs may also increase the opportunity for sperm to reach storage sites. Although the exact function of copulatory plugs is often unclear, they appear to be important in sperm competition. For example, in primates there is a positive relationship between the degree of seminal coagulation and female multiple mating [89], suggesting that sperm competition selects for the formation of copulatory plugs. Similarly, copulatory plugs are larger in promiscuous rodents than in monogamous species [80].

Copulatory plugs are found in many bat species (Table **3**). By determining if these plugs are male or female in origin, we may gain some insight into their function. This is because copulatory plugs produced by females are likely to prevent sperm leakage, while plugs produced by males may be more likely to inhibit re-mating or prevent sperm dumping [79]. Unfortunately, for many bat species we do not yet know if the plugs observed are of male or female origin (Table **3**). Furthermore, we do not yet know if the presence of copulatory plugs is associated with a high risk of sperm competition. However, many of the species that do have copulatory plugs, are those where females store sperm for prolonged periods of time and so, there is the potential for sperm competition (Table **3**). Clearly, there is an awful lot we do not know about the function of these plugs in bats but equally, tremendous potential for us to improve our understanding.

Table 3. Summarizing representative data on bat copulatory plug presence [1] or absence [0] and the sex responsible for their production (origin) either; female (F) or male (M). Also illustrated is the presence [1] or absence [0] of long-term female sperm storage (delayed fertilization, delayed ovulation) or lack of long-term sperm storage [0]. A few other reproductive states are illustrated DI=delayed implantation and DD=delayed development. Dashes indicate unknown values. Mating system assignments follow those of Wilkinson and McCracken [36]: (MM (multiple-male), SM (single-male), MF (multiple-female), SF (single-female). Table is based in part on the excellent summary tables of Fenton [107] and Wilkinson and McCracken [36]. Data presented by Family.

Genus	**Species**	**Common Name**	**Plug**	**Plug Origin**	**Female Sperm Storage**	**Mating System**	**Ref.**
Hipposideridae							
Hipposideros	*caffer*	Sundevall's leaf-nosed bat	1	F	-	-	[123 - 125]
Rhinonicteris	*aurantius*	orange leaf-nosed bat	0,1	-	-	-	[101]
Molossidae							
Molossus	*ater (rufus)*	black mastiff bat	1	-	1?	-	[126]
Mormopterus	*planiceps*	little mastiff bat	1	-	1	-	[127]
Tadarida	*brasiliensis*	Mexican free-tailed bat	1	-	0	MM-MF	[36, 52]
Mystaacinidae							
Mystacina	*tuberculata.*	short-tailed bats	1	M	-		[128]
Phyllostomidae							
Carollia	*perspicillata*	Seba's short-tailed fruit bat	1	-	0 (DD)	SM-MF	[36, 129, 130]
Leptonycteris	*yerbabuenae*	lesser long-nosed bat	1	-			[131]
Pteropodidae							

(Table 3) contd.....

Genus	Species	Common Name	Plug	Plug Origin	Female Sperm Storage	Mating System	Ref.
Pteropus	*poliocephalus*	grey-headed flying fox	1	M	-	MM-MF	[36, 132]
Rhinolophidae							
Rhinolophids	*(various sp.)*	horseshoe bats	1	F	1	-	[133, 134]
Rhinolophus	*capensis*	Cape horseshoe bat	1	F	-	-	[124]
Rhinolophus	*ferrumequinum*	greater horseshoe bat	1	M/F	1	SM-MF	[36, 134]; [135]; [137]
Rhinolophus	*hipposiderous*	lesser horseshoe bat	1	M	1	-	[134, 138, 139]
Rhinolophus	*megaphyllus*	Eastern horseshoe bat	0	-	-	-	[140]
Vespertilionidae							
Chalinolobus	*gouldii*	Gould's wattled bat	1	F	1	-	[141]
Chalinolobus	*morio*	wattled bats	1	-	-	-	[142]
Corynorhinus	*townsendii*	Townsend's big-eared bat	0	-	1?, DI	-	[143]
Eptesicus	*fuscus*	big brown bat	1	F	1	-	[144], [145]
Eptesicus	*regulus*	Southern forest bat	1	F	1	-	[146]
Eptesicus[†]	*vulturnus*	little forest bat	1	-	1	-	[147, 148]
Eptesicus[†]	*regulus*	Southern forest bat	1	-	1	-	[147, 148]
Eptesicus[†]	*darlingtoni*	large forest bat	1	-	1	-	[147, 148]
Miniopterus	schreibersii	Japanese long-fingered bat	1	-	Limited	-	[81, 149]
Myotis	*lucifugus*	little brown bat	0	-	1	MM-MF	[36, 46, 145, 150]
Myotis	*ricketti*	Rickett's big-earred bat	0	-	1		[151]
Nyctalus	*noctula*	common noctule	1	F	1	SM-MF	[36, 46, 152]
Nyctophilus	*geoffroyi*	Australian long-eared bat	1	-	1	-	[36, 153]
Nyctophilus	*gouldi*	Gould's long-eared bat	1	-	1	-	[108]

(Table 3) contd.....

Genus	Species	Common Name	Plug	Plug Origin	Female Sperm Storage	Mating System	Ref.
Pipistrellus	*abramus*	Japanese house bat	0	-	1	-	[81, 154]
Pipistrellus	*nanus*	banana pipistrelle	0	-	1	-	[48, 155]
Pipistrellus	*pipistrellus*	common pipistrelle	1	F	1	SM-MF	[36, 46, 156, 157]
Plecotus	*auritus*	brown long-eared bat	0?	-	1	MM-MF	[36, 158, 159]
Scotophilus	*heathi*	greater Asiatic yellow bat	1	-	1	-	[160]

Genital Morphology

The underlying structure of the penis is conserved across the mammals: a fleshy cylinder surrounds a central hydraulic skeleton, which allows the penis to become erect when filled with fluid [90]. In many mammals, a central bone, the baculum, is present inside this cylinder [95]. Beyond these central features, penis morphology varies enormously, even between species that otherwise closely resemble one another [91, 92]. This variation may be driven by both sperm competition and cryptic female choice. Female choice may select for the genital morphology that best stimulates females during mating. In turn, females may benefit from this *via* increased fertility or by producing sons with the ability to also stimulate females (Fisherian sexual selection). Alternatively, sperm competition might favour penis morphology that help males remove the sperm of their rival males or displace copulatory plugs [93], and there are many examples of penis morphology evolving in response to the risk of sperm competition. For example, in primates over half of taxa with multimale-multifemale mating systems have elongated penes, while only 8% of monogamous species do [94].

In bats, penis morphology varies enormously across species [95, 96]. So much so, that penis morphology is used as species identification tool [97]. Given that male bats may use their penis to remove rival's copulatory plugs [52], it seems likely that genital variation is driven in part, by sperm competition. Most research relating sperm competition to penis morphology has focused on the baculum. The shape and size of bacula (os penes) shows great variety amongst bats [95, 98]. It is also functionally important; 3D modelling shows that in the common noctule bat *Nyctalus noctula*, the common pipistrelle *Pipistrellus pipistrellus,* and Nathusius' pipistrelle *Pipistrellus nathusii,* the baculum protects the urethral opening and helps provide stiffness within the glans [96]. Comparative analyses show that at the species level, baculum length is positively correlated with relative testis mass and the interaction between testis mass and mating system [95]. However, after controlling for phylogeny, these patterns disappear, suggesting that evolutionary history rather than similar selection caused these associations [95]. This also suggests that baculum length is not sexually selected, which is perhaps not surprising as the bone is fully enclosed in tissue. This result is consistent with another comparative analysis across four mammalian orders (bats, carnivores, primates, and rodents), which found that while baculum and glans penis length positively correlated with testis mass in rodents and carnivores, this was not the case in bats or primates [99].

Bat penises also vary in their external structure. In some species, penises are covered in small spines [100, 101], while in other species the glans penes are large and elaborate [102], possibly functioning as a scoop to remove stored sperm. Unfortunately, we lack formal analysis relating this morphology to the risk of sperm competition. However, this elaboration seems to be present in species with delayed fertilization (*e.g. Lasiurus cinereus*) and absent in species that lack this reproductive delay (*e.g. Carollia perspicillata*) [49]. If this is a consistent trend and penis elaborations play a role in sperm competition, it is unclear exactly what role these elaborations may play. Penile spines may not reach the region of the female where sperm are stored, making their presence harder to explain in light of sperm competition. Orr and Zuk [49] suggested that perhaps the enlarged glans penes pulls trapped sperm down to the region of the female where male spines penetrate, or perhaps they stimulate sperm dumping by females as seen in other taxa [103]. In other mammals such as cats, penile spines are thought to elicit ovulation- however in many bats the physiology is reversed (delayed ovulation re-sperm storage). However, the pervasive selective agent responsible for penis evolution appears to be female choice [91, 92], so this seems likely to be case for bat penises too [161].

Post-copulatory sexual selection may also shape the evolution of female genitalia. By having more elaborate or long oviducts, sperm need to travel greater distances to fertilize ova. By increasing the length or complexity of this gauntlet that sperm need to run, female form could select for high quality sperm [12, 104]. In support of this idea, a comparative analyses of 48 species across 33 mammalian genera found that oviduct length is positively correlated with relative testis size and that oviduct length also tightly correlated with sperm mid-piece volume [70]. Data for the two species of bats included in these analyses suggest that bats conform to these general trends [70]. Unfortunately, we lack the data needed to really understand how female genitals have evolved in response to post-copulatory sexual selection and we do not have any evidence relating sperm performance to either male performance or the quality of the haploid genotype deliver by the sperm for any bat.

Behavioural Adaptations

Behaviour and post-copulatory sexual selection are clearly related, with sexual behaviours influencing variance in fertilization success and the potential for post-copulatory sexual selection can affect sexual behaviour. For example, if females terminate a mating early, then it may be female receptivity, rather than testis volume and sperm numbers, that ultimately decides how many sperm a male transfers at mating. In these cases males could improve success by encouraging females to mate for longer. There is some evidence for males engaging in behaviours that increase mating duration in bats. In the Indian flying-fox *Pteropus giganteus* males engage in cunnilingus both before and after mating. When a male approaches a receptive female, he licks her vagina for around 50 seconds. This is typically followed by mating, after which a male resumes cunnilingus for up to three minutes [105]. How long a male engages in pre-copulatory cunnilingus is positively correlated with how long he mates. In theory, this longer mating duration could increase the time for sperm to transfer to females or increase the likelihood females favour a male's sperm *via* cryptic female choice, and therefore, improve a male's siring success. Females may also engage in oral sex in some bats. In the greater short-nosed fruit bat *Cynopterus sphinx*, females perform fellatio prior to copulation [106]. This may be a means by which females assess a mate's quality and subsequently, her willingness to mate or the likelihood that she facilitates or inhibits that male's sperm from fertilizing her ova. Perhaps more prosaically this is a way to clean the intromittent organ and reduce disease transmission? This is an area that clearly warrants further attention. It is interesting that of the behaviours described above all are from Old World fruit bats. It is highly probable that this is not because these behaviours are restricted to this group but rather because these species are more diurnal and conspicuous and thus more easily observed relative to smaller cave-dwelling and nocturnal non-pteropodid bats.

Cryptic Female Choice

It is often impossible to completely disentangle sperm competition and cryptic female choice as processes and hence their influences on reproductive behaviours and characters must also be intermeshed [13]. Many of the traits discussed above will therefore be driven by a combination of the two processes, for example, females may prefer males with larger sperm and these may also have a competitive advantage in sperm competition. Even when these mechanisms are theoretically quite distinct, empirically it is challenging to tell them apart. However, there are clear signs that female bats exert cryptic choice.

First, data on two bat species suggests that multiple mating is associated with the evolution of long convoluted oviducts [70]. Thus tract morphology can be thought of as a morphological trait that enables cryptic female choice [22]. While we do not know if females have other physiological mechanisms to bias fertilization success, it is easy to imagine that they do. For example, the female sperm stores in little brown bats contain high levels of Zn, which can help promote sperm function and survival. Some of this Zn is female in origin and may have been sequestered from other tissues [83]. By differentially sequestering and storing Zn, females may be able to bias fertilization towards particular mates. While this idea is speculative, understanding female adaptations to improve the fertilization success of particular sperm is an exciting avenue of future research.

There are also female behaviours that may bias post-copulatory fertilization success in favour of particular mates. For example, in the Indian flying-fox (*P. giganteus*) females mate for longer with males that engage in pre-copulatory cunnilingus [105]. This will probably improve that male's fertilization success and so is an example of a behavioural trait to exert cryptic choice. Female bats have been seen to remove copulatory plugs and then dump the sperm from a previous mating [107, 108]. This clearly severely reduces the fertilization success of her most recent mate and so is another behavior that biases a particular male's fertilization odds.

FUTURE DIRECTIONS AND CHALLENGES

Some elements of post-copulatory sexual selection are well established in bats – sperm competition selects for larger testes for example – but as is clear from the preceding discussion much is still unknown. This is partly because of the difficulty in working with a volant and nocturnal subject, but also because there are just so many bat species that it is impossible to know what is yet to be discovered. Additionally, because sample sizes in experimental studies of bat reproduction are often extremely small (*e.g.*, [57]) it will be difficult to detect effects that are small and the extreme disconnect between sperm production, mating and birth also makes it challenging to link male and sperm phenotype with siring success. Nonetheless, the range of topics open to further work is broad. For example, how does penile morphology affect paternity, do taxa with females sperm storage have sperm attributes that differ from non-storing species, how do females store viable sperm for so long and does it aid choice of sires, do species that mate well after the cessation of spermatogenesis have higher quality sperm, and so on. Future research into the extraordinary reproductive biology of this broad taxon will surely provide novel insights into cryptic female choice and sperm biology, even if only to show how it is possible to store sperm for such long periods. There is certainly plenty to do for future researcher with the tenacity to focus on bats.

CONFLICT OF INTEREST

The authors declare no conflict of interest, financial or otherwise.

ACKNOWLEDGEMENTS

None declared

REFERENCES

[1] Andersson M. Sexual selection: Monographs in behavior and ecology. Princeton, NJ: Princeton Univ. Press 1994.

[2] Hosken DJ, Alonzo SH, Wedell N. Why aren't signals of female quality more common? Anim Behav 2016; 114: 199-201.
[http://dx.doi.org/10.1016/j.anbehav.2016.02.015]

[3] Darwin C. The descent of man, and selection in relation to sex. John Murray & Sons 1872.

[4] Trivers R. Parental investment and sexual selection. Sexual Selection & the Descent of Man. New York: Aldine de Gruyter 1972.

[5] Bateman AJ. Intra-sexual selection in *Drosophila*. Heredity 1948; 2(3): 349-68.
[http://dx.doi.org/10.1038/hdy.1948.21] [PMID: 18103134]

[6] Hosken DJ, House CM. Sexual selection. Curr Biol 2011; 21(2): R62-5.
[http://dx.doi.org/10.1016/j.cub.2010.11.053] [PMID: 21256434]

[7] Rosenqvist G. Male mate choice and female-female competition for mates in the pipefish *Nerophis ophidion*. Anim Behav 1990; 39(6): 1110-5.
[http://dx.doi.org/10.1016/S0003-3472(05)80783-3]

[8] Hosken DJ, Stockley P. Benefits of polyandry: a life history perspective. Evolutionary Biology. Springer 2003; pp. 173-94.
[http://dx.doi.org/10 1007/978-1-4757-5190-1_4]

[9] Taylor ML, Price TA, Wedell N. Polyandry in nature: a global analysis. Trends Ecol Evol 2014; 29(7): 376-83.
[http://dx.doi.org/10.1016/j.tree.2014.04.005] [PMID: 24831458]

[10] Baer B, Eubel H, Taylor NL, O'Toole N, Millar AH. Insights into female sperm storage from the spermathecal fluid proteome of the honeybee *Apis mellifera*. Genome Biol 2009; 10(6): R67.
[http://dx.doi.org/10.1186/gb-2009-10-6-r67] [PMID: 19538722]

[11] Parker GA. Sperm competition and its evolutionary consequences in the insects. Biol Rev Camb Philos Soc 1970; 45(4): 525-67.
[http://dx.doi.org/10.1111/j.1469-185X.1970.tb01176.x]

[12] Eberhard WG. Female control: sexual selection by cryptic female choice. Princeton University Press 1996.

[13] Pitnick S, Hosken DJ. Postcopulatory sexual selection. Evol Behav Ecol Oxford University Press, Oxford 2010; 379-99.

[14] Parker GA. Sperm competition games: raffles and roles. Proc Biol Sci 1990; 242(1304): 120-6.
[http://dx.doi.org/10.1098/rspb.1990.0114]

[15] Holman L, Stürup M, Trontti K, Boomsma JJ. Random sperm use and genetic effects on worker caste fate in *Atta colombica* leaf-cutting ants. Mol Ecol 2011; 20(23): 5092-102.
[http://dx.doi.org/10.1111/j.1365-294X.2011.05338.x] [PMID: 22053996]

[16] Hosken DJ. Sperm displacement in yellow dung flies: a role for females. Trends Ecol Evol 1999; 14(7): 251-2.
[http://dx.doi.org/10.1016/S0169-5347(99)01641-9] [PMID: 10370258]

[17] Simmons LW, Parker GA, Stockley P. Sperm displacement in the yellow dung fly, *Scatophaga stercoraria*: an investigation of male and female processes. Am Nat 1999; 153(3): 302-14.
[http://dx.doi.org/10.1086/303171]

[18] Snook RR. Sperm in competition: not playing by the numbers. Trends Ecol Evol 2005; 20(1): 46-53.
[http://dx.doi.org/10.1016/j.tree.2004.10.011] [PMID: 16701340]

[19] Watson PJ. Multiple paternity and first mate sperm precedence in the sierra dome spider, *Linyphia litigiosa* Keyserling (Linyphiidae). Anim Behav 1991; 41(2): 135-48.
[http://dx.doi.org/10.1016/S0003-3472(05)80509-3]

[20] Thornhill R. Cryptic female choice and its implications in the scorpionfly *Harpobittacus nigriceps*. Am Nat 1983; 122(6): 765-88.
[http://dx.doi.org/10.1086/284170]

[21] Hodgson DJ, Hosken DJ. Sperm competition promotes the exploitation of rival ejaculates. J Theor Biol 2006; 243(2): 230-4.
[http://dx.doi.org/10.1016/j.jtbi.2006.06.024] [PMID: 16901507]

[22] Arnqvist G. Cryptic female choice. In: Shuker D, Simmons L, Eds. The Evolution of Insect Mating Systems. Oxford University Press 2014; pp. 204-20.
[http://dx.doi.org/10.1093/acprof:oso/9780199678020.003.0011]

[23] Rosengrave P, Gemmell NJ, Metcalf V, McBride K, Montgomerie R. A mechanism for cryptic female choice in chinook salmon. Behav Ecol 2008; 19(6): 1179-85.
[http://dx.doi.org/10.1093/beheco/arn089]

[24] Pizzari T, Birkhead TR. Female feral fowl eject sperm of subdominant males. Nature 2000; 405(6788): 787-9.
[http://dx.doi.org/10.1038/35015558] [PMID: 10866198]

[25] Manier MK, Belote JM, Berben KS, *et al.* Rapid diversification of sperm precedence traits and processes among three sibling *Drosophila* species. Evolution 2013; 67(8): 2348-62.
[http://dx.doi.org/10.1111/evo.12117] [PMID: 23888856]

[26] Manier MK, Lüpold S, Pitnick S, Starmer WT. An analytical framework for estimating fertilization bias and the fertilization set from multiple sperm-storage organs. Am Nat 2013; 182(4): 552-61.
[http://dx.doi.org/10.1086/671782] [PMID: 24021407]

[27] Zeh JA, Zeh DW. Reproductive mode and the genetic benefits of polyandry. Anim Behav 2001; 61(6): 1051-63.
[http://dx.doi.org/10.1006/anbe.2000.1705]

[28] Hosken DJ, Blanckenhorn WU. Female multiple mating, inbreeding avoidance, and fitness: it is not only the magnitude of costs and benefits that counts. Behav Ecol 1999; 10(4): 462-4.
[http://dx.doi.org/10.1093/beheco/10.4.462]

[29] Tregenza T, Wedell N. Polyandrous females avoid costs of inbreeding. Nature 2002; 415(6867): 71-3.
[http://dx.doi.org/10.1038/415071a] [PMID: 11780118]

[30] Slatyer RA, Mautz BS, Backwell PR, Jennions MD. Estimating genetic benefits of polyandry from experimental studies: a meta-analysis. Biol Rev Camb Philos Soc 2012; 87(1): 1-33.
[http://dx.doi.org/10.1111/j.1469-185X.2011.00182.x] [PMID: 21545390]

[31] Birkhead TR, Møller AP. Sperm competition and sexual selection. Academic Press 1998.

[32] Simmons LW. Sperm competition and its evolutionary consequences in the insects. Princeton University Press 2001.

[33] Racey PA. Ecology of bat reproduction. Ecology of bats. Springer 1982; pp. 57-104. [http://dx.doi.org/10.1007/978-1-4613-3421-7_2]

[34] Hosken DJ. Reproduction and the female reproductive cycle of *Nyctophilus geoffroyi* and *N. major* (Chiroptera: Vespertilionidae) from South-western Australia. Aust J Zool 1997; 45(5): 489-504. [http://dx.doi.org/10.1071/ZO96060]

[35] McCracken GF, Wilkinson GS. Bat mating systems. In: Crichton EG, Krutzsch PH, Eds. Reproductive Biology of Bats. Academics Press 2000; pp. 3-5. [http://dx.doi.org/10.1016/B978-012195670-7/50009-6]

[36] Wilkinson GS, McCracken GF. Bats and balls: sexual selection and sperm competition in the Chiroptera. In: Kunz TH, Fenton MB, Eds. Bat Ecology. Chicago, Illinois: University of Chicago Press 2003; pp. 128-55.

[37] Wai-Ping V, Fenton MB. Nonselective mating in little brown bats (*Myotis lucifugus*). Wildl Monogr 1988; 35(3): 1-60.

[38] Hosken DJ. Sperm competition in bats. Proc Biol Sci 1997; 264(1380): 385-92. [http://dx.doi.org/10.1098/rspb.1997.0055] [PMID: 9107054]

[39] Kunz TH, Kurta A. Size of bats at birth and maternal investment during pregnancy. Symposia of the Zoological Society of London 1987; 57: 79-106.

[40] Vonhof MJ, Barber D, Fenton MB, Strobeck C. A tale of two siblings: multiple paternity in big brown bats (*Eptesicus fuscus*) demonstrated using microsatellite markers. Mol Ecol 2006; 15(1): 241-7. [http://dx.doi.org/10.1111/j.1365-294X.2005.02801.x] [PMID: 16367843]

[41] Fox S, Spencer H, O'Brien GM. Analysis of twinning in flying-foxes (Megachiroptera) reveals superfoetation and multiple-paternity. Acta Chiropt 2008; 10(2): 271-8. [http://dx.doi.org/10.3161/150811008X414845]

[42] Wimsatt WA. Further studies on the survival of spermatozoa in the female reproductive tract of the bat. Anat Rec 1944; 88(2): 193-204. [http://dx.doi.org/10.1002/ar.1090880206]

[43] Birkhead TR, Møller AP. Sexual selection and the temporal separation of reproductive events: sperm storage data from reptiles, birds and mammals. Biol J Linn Soc Lond 1993; 50(4): 295-311. [http://dx.doi.org/10.1111/j.1095-8312.1993.tb00933.x]

[44] Hosken DJ, O'Shea JE, Blackberry MA. Blood plasma concentrations of progesterone, sperm storage and sperm viability and fertility in Gould's wattled bat (*Chalinolobus gouldii*). J Reprod Fertil 1996; 108(2): 171-7. [http://dx.doi.org/10.1530/jrf.0.1080171] [PMID: 9038773]

[45] Holt WV. Mechanisms of sperm storage in the female reproductive tract: an interspecies comparison. Reprod Domest Anim 2011; 46 (Suppl. 2): 68-74. [http://dx.doi.org/10.1111/j.1439-0531.2011.01862.x] [PMID: 21884282]

[46] Racey PA. The prolonged storage and survival of spermatozoa in Chiroptera. J Reprod Fertil 1979; 56(1): 391-402. [http://dx.doi.org/10.1530/jrf.0.0560391] [PMID: 381655]

[47] Pfeiffer B, Mayer F. Spermatogenesis, sperm storage and reproductive timing in bats. J Zool 2013; 289(2): 77-85. [http://dx.doi.org/10.1111/j.1469-7998.2012.00970.x]

[48] Bernard RTF, Cumming GS. African bats: evolution of reproductive patterns and delays. Q Rev Biol 1997; 72(3): 253-74. [http://dx.doi.org/10.1086/419859] [PMID: 9293029]

[49] Orr TJ, Zuk M. Reproductive delays in mammals: an unexplored avenue for post-copulatory sexual selection. Biol Rev Camb Philos Soc 2014; 89(4): 889-912.

[http://dx.doi.org/10.1111/brv.12085] [PMID: 24517909]

[50] Crichton EG. Sperm storage and fertilization. In: Crichton EG, Krutzsch PH, Eds. Reproductive Biology of Bats. Elsevier BV 2000; pp. 295-320. [http://dx.doi.org/10.1016/B978-012195670-7/50008-4]

[51] Hosken DJ, Blackberry MA, Stewart TB, Stucki AF. The male reproductive cycle of three species of Australian vespertilionid bat. J Zool 1998; 245(3): 261-70. [http://dx.doi.org/10.1111/j.1469-7998.1998.tb00100.x]

[52] Keeley AH, Keeley BW. The mating system of *Tadarida brasiliensis* (Chiroptera: Molossidae) in a large highway bridge colony. J Mammal 2004; 85(1): 113-9. [http://dx.doi.org/10.1644/BME-004]

[53] Parker GA. The evolution of expenditure on testes. J Zool 2016; 298: 3-19. [http://dx.doi.org/10.1111/jzo.12297]

[54] Hosken DJ, Ward PI. Experimental evidence for testis size evolution via sperm competition. Ecol Lett 2001; 4(1): 10-3. [http://dx.doi.org/10.1046/j.1461-0248.2001.00198.x]

[55] Gage MJG. Associations between body size, mating pattern, testis size and sperm lengths across butterflies. Proc Biol Sci 1994; 258(1353): 247-54. [http://dx.doi.org/10.1098/rspb.1994.0169]

[56] Stockley P, Gage MJ, Parker GA, Møller AP. Sperm competition in fishes: the evolution of testis size and ejaculate characteristics. Am Nat 1997; 149(5): 933-54. [http://dx.doi.org/10.1086/286031] [PMID: 18811256]

[57] Hosken DJ. Testes mass in megachiropteran bats varies in accordance with sperm competition theory. Behav Ecol Sociobiol 1998; 44(3): 169-77. [http://dx.doi.org/10.1007/s002650050529]

[58] Orr TJ, Zuk M. Does delayed fertilization facilitate sperm competition in bats? Behav Ecol Sociobiol 2013; 67(12): 1903-13. [http://dx.doi.org/10.1007/s00265-013-1598-2]

[59] Pitnick S, Hosken DJ, Birkhead TR. Birkhead TR, Hosken DJ, Pitnick S. Sperm morphological diversity. Sperm Biology: An Evolutionary Perspective. Academics Press 2008; pp. 69-149.

[60] Racey PA, Potts DM. Relationship between stored spermatozoa and the uterine epithelium in the pipistrelle bat (*Pipistrellus pipistrellus*). J Reprod Fertil 1970; 22(1): 57-63. [http://dx.doi.org/10.1530/jrf.0.0220057] [PMID: 4192709]

[61] Pitnick S. Investment in testes and the cost of making long sperm in *Drosophila*. Am Nat 1996; 148(1): 57-80. [http://dx.doi.org/10.1086/285911]

[62] Kleven O, Fossøy F, Laskemoen T, Robertson RJ, Rudolfsen G, Lifjeld JT. Comparative evidence for the evolution of sperm swimming speed by sperm competition and female sperm storage duration in passerine birds. Evolution 2009; 63(9): 2466-73. [http://dx.doi.org/10.1111/j.1558-5646.2009.00725.x] [PMID: 19453726]

[63] Birkhead TR, Martínez JG, Burke T, Froman DP. Sperm mobility determines the outcome of sperm competition in the domestic fowl. Proc Biol Sci 1999; 266(1430): 1759-64. [http://dx.doi.org/10.1098/rspb.1999.0843] [PMID: 10577160]

[64] Malo AF, Garde JJ, Soler AJ, García AJ, Gomendio M, Roldan ER. Male fertility in natural populations of red deer is determined by sperm velocity and the proportion of normal spermatozoa. Biol Reprod 2005; 72(4): 822-9. [http://dx.doi.org/10.1095/biolreprod.104.036368] [PMID: 15576823]

[65] Gomendio M, Roldan ER. Implications of diversity in sperm size and function for sperm competition

and fertility. Int J Dev Biol 2008; 52(5-6): 439-47.
[http://dx.doi.org/10.1387/ijdb.082595mg] [PMID: 18649256]

[66] Fitzpatrick JL, Montgomerie R, Desjardins JK, Stiver KA, Kolm N, Balshine S. Female promiscuity promotes the evolution of faster sperm in cichlid fishes. Proc Natl Acad Sci USA 2009; 106(4): 1128-32.
[http://dx.doi.org/10.1073/pnas.0809990106] [PMID: 19164576]

[67] Gomendio M, Roldan ER. Sperm competition influences sperm size in mammals. Proc Biol Sci 1991; 243(1308): 181-5.
[http://dx.doi.org/10.1098/rspb.1991.0029] [PMID: 1675796]

[68] Malo AF, Gomendio M, Garde J, Lang-Lenton B, Soler AJ, Roldan ER. Sperm design and sperm function. Biol Lett 2006; 2(2): 246-9.
[http://dx.doi.org/10.1098/rsbl.2006.0449] [PMID: 17148374]

[69] Humphries S, Evans JP, Simmons LW. Sperm competition: linking form to function. BMC Evol Biol 2008; 8(1): 319.
[http://dx.doi.org/10.1186/1471-2148-8-319] [PMID: 19032741]

[70] Anderson MJ, Dixson AS, Dixson AF. Mammalian sperm and oviducts are sexually selected: evidence for co-evolution. J Zool 2006; 270(4): 682-6.
[http://dx.doi.org/10.1111/j.1469-7998.2006.00173.x]

[71] Gage MJG, Freckleton RP. Relative testis size and sperm morphometry across mammals: no evidence for an association between sperm competition and sperm length. Proc Biol Sci 2003; 270(1515): 625-32.
[http://dx.doi.org/10.1098/rspb.2002.2258] [PMID: 12769463]

[72] Fasel NJ, Helfenstein F, Buff S, Richner H. Electroejaculation and semen buffer evaluation in the microbat *Carollia perspicillata*. Theriogenology 2015; 83(5): 904-10.
[http://dx.doi.org/10.1016/j.theriogenology.2014.11.030] [PMID: 25541424]

[73] Martin PA, Reimers TJ, Lodge JR, Dziuk PJ. The effect of ratios and numbers of spermatozoa mixed from two males on proportions of offspring. J Reprod Fertil 1974; 39(2): 251-8.
[http://dx.doi.org/10.1530/jrf.0.0390251] [PMID: 4416213]

[74] Poiani A. Complexity of seminal fluid: a review. Behav Ecol Sociobiol 2006; 60(3): 289-310.
[http://dx.doi.org/10.1007/s00265-006-0178-0]

[75] Peitz B. Effects of seminal vesicle fluid components on sperm motility in the house mouse. J Reprod Fertil 1988; 83(1): 169-76.
[http://dx.doi.org/10.1530/jrf.0.0830169] [PMID: 3397934]

[76] Herndon LA, Wolfner MF. A *Drosophila* seminal fluid protein, Acp26Aa, stimulates egg laying in females for 1 day after mating. Proc Natl Acad Sci USA 1995; 92(22): 10114-8.
[http://dx.doi.org/10.1073/pnas.92.22.10114] [PMID: 7479736]

[77] Kubli E. The *Drosophila* sex-peptide: a peptide pheromone involved in reproduction. Adv Develop Biochem 1996; 4: 99-128.
[http://dx.doi.org/10.1016/S1064-2722(08)60058-4]

[78] Radhakrishnan P, Taylor PW. Seminal fluids mediate sexual inhibition and short copula duration in mated female Queensland fruit flies. J Insect Physiol 2007; 53(7): 741-5.
[http://dx.doi.org/10.1016/j.jinsphys.2006.10.009] [PMID: 17187820]

[79] Voss R. Male accessory glands and the evolution of copulatory plugs in rodents. Occasional papers of the Museum of Zoology, University of Michigan 1979.

[80] Ramm SA, Parker GA, Stockley P. Sperm competition and the evolution of male reproductive anatomy in rodents. Proc Biol Sci 2005; 272(1566): 949-55.
[http://dx.doi.org/10.1098/rspb.2004.3048] [PMID: 16024351]

[81] Mōri T, Uchida TA. Sperm storage in the reproductive tract of the female Japanese long-fingered bat, *Miniopterus schreibersii fuliginosus*. J Reprod Fertil 1980; 58(2): 429-33. [http://dx.doi.org/10.1530/jrf.0.0580429] [PMID: 7191895]

[82] Hunter AG, Johnson WL, Barker LDS, Fahning ML, Schultz RH. Bat seminal vesicle protein: its characterization and physiological properties. J Reprod Fertil 1971; 24(2): 179-86. [http://dx.doi.org/10.1530/jrf.0.0240179] [PMID: 5102534]

[83] Crichton EG, Krutzsch PH, Chvapil M. Studies on prolonged spermatozoa survival in Chiroptera—II. The role of zinc in the spermatozoa storage phenomenon. Comp Biochem Physiol A 1982; 71: 71-7. [http://dx.doi.org/10.1016/0300-9629(82)90369-3]

[84] Miki K. Energy metabolism and sperm function. Soc Reprod Fertil Suppl 2007; 65: 309-25. [PMID: 17644971]

[85] Crichton EG, Krutzsch PH, Wimsatt WA. Studies on prolonged spermatozoa survival in Chiroptera-I. The role of uterine free fructose in the spermatozoa storage phenomenon. Comp Biochem Physiol A 1981; 70(3): 387-95. [http://dx.doi.org/10.1002/aja.1001650406] [PMID: 7158612]

[86] Engle ET. The copulation plug and the accessory genital glands of mammals. J Mammal 1926; 7(2): 119-26. [http://dx.doi.org/10.2307/1373679]

[87] Martan J, Shepherd BA. The role of the copulatory plug in reproduction of the guinea pig. J Exp Zool 1976; 196(1): 79-84. [http://dx.doi.org/10.1002/jez.1401960108] [PMID: 932657]

[88] Dewsbury DA. A test of the role of copulatory plugs in sperm competition in deer mice (*Peromyscus maniculatus*). J Mammal 1988; 69(4): 854-7. [http://dx.doi.org/10.2307/1381648]

[89] Dixson AL, Anderson MJ. Sexual selection, seminal coagulation and copulatory plug formation in primates. Folia Primatol 2002; 73(2-3): 63-9. [http://dx.doi.org/10.1159/000064784] [PMID: 12207054]

[90] Kelly DA. Turtle and mammal penis designs are anatomically convergent. Proc Biol Sci 2004; 271 (Suppl. 5): S293-5. [http://dx.doi.org/10.1098/rsbl.2004.0161] [PMID: 15503998]

[91] Eberhard WG. Animal genitalia and female choice. Am Sci 1990; 78(2): 134-41.

[92] Hosken DJ, Stockley P. Sexual selection and genital evolution. Trends Ecol Evol 2004; 19(2): 87-93. [http://dx.doi.org/10.1016/j.tree.2003.11.012] [PMID: 16701234]

[93] Ginsberg JR, Huck UW. Sperm competition in mammals. Trends Ecol Evol 1989; 4(3): 74-9. [http://dx.doi.org/10.1016/0169-5347(89)90152-3] [PMID: 21227319]

[94] Dixson A. Sexual selection by cryptic female choice and the evolution of primate sexuality. Evol Anthropol 2002; 11(S1): 195-9. [http://dx.doi.org/10.1002/evan.10090]

[95] Hosken DJ, Jones K, Chipperfield K, Dixson A. Is the bat os penis sexually selected? Behav Ecol Sociobiol 2001; 50(5): 450-60. [http://dx.doi.org/10.1007/s002650100389]

[96] Herdina AN, Kelly DA, Jahelková H, Lina PH, Horáček I, Metscher BD. Testing hypotheses of bat baculum function with 3D models derived from microCT. J Anat 2015; 226(3): 229-35. [http://dx.doi.org/10.1111/joa.12274] [PMID: 25655647]

[97] Krutzsch PH, Vaughan TA. Additional data on the bacula of North American bats. J Mammal 1955; 36(1): 96-100. [http://dx.doi.org/10.2307/1375727]

[98] Krutzsch PH. Variation in the os penis of tropical fruit bats. J Mammal 1959; 40(3): 387-92. [http://dx.doi.org/10.2307/1376563]

[99] Ramm SA. Sexual selection and genital evolution in mammals: a phylogenetic analysis of baculum length. Am Nat 2007; 169(3): 360-9. [http://dx.doi.org/10.1086/510688] [PMID: 17238128]

[100] Ryan JM. Comparative morphology of the glans-penis in Molossus, Promops, and Eumops (Chiroptera, Molossidae). Bull Am Mus Nat Hist 1991; 206: 122-37.

[101] Armstrong KN. A description and discussion of the penile morphology of *Rhinonicteris aurantius* (Gray, 1845) (Microchiroptera: Hipposideridae). Aust Mammal 2005; 27(2): 161-7.

[102] Matthews LH. Notes on the genitalia and reproduction of some African rats. Proc Zool Soc Lond. Wiley Online Library 1942; B111: pp. 289-342. [http://dx.doi.org/10.1111/j.1469-7998.1942.tb00053.x]

[103] Snook RR, Hosken DJ. Sperm death and dumping in *Drosophila*. Nature 2004; 428(6986): 939-41. [http://dx.doi.org/10.1038/nature02455] [PMID: 15118726]

[104] Gomendio M, Roldan ER. Coevolution between male ejaculates and female reproductive biology in eutherian mammals. Proc Biol Sci 1993; 252(1333): 7-12. [http://dx.doi.org/10.1098/rspb.1993.0039] [PMID: 8389048]

[105] Maruthupandian J, Marimuthu G. Cunnilingus apparently increases duration of copulation in the Indian flying fox, *Pteropus giganteus*. PLoS One 2013; 8(3): e59743. [http://dx.doi.org/10.1371/journal.pone.0059743] [PMID: 23544092]

[106] Tan M, Jones G, Zhu G, *et al.* Fellatio by fruit bats prolongs copulation time. PLoS One 2009; 4(10): e7595. [http://dx.doi.org/10.1371/journal.pone.0007595] [PMID: 19862320]

[107] Fenton MB. Sperm competition? The case of vespertilionid and rhinolophid bats. Sperm Competition and the Evolution of Animal Mating Systems. 1984; p. 573.587.

[108] Phillips WR, Inwards SJ. The annual activity and breeding cycles of Gould's long-eared bat, *Nyctophilus gouldi* (Microchiroptera: Vespertiionidae). Aust J Zool 1985; 33(2): 111-26. [http://dx.doi.org/10.1071/ZO9850111]

[109] Roellig K, Menzies BR, Hildebrandt TB, Goeritz F. The concept of superfetation: a critical review on a 'myth' in mammalian reproduction. Biol Rev Camb Philos Soc 2011; 86(1): 77-95. [http://dx.doi.org/10.1111/j.1469-185X.2010.00135.x] [PMID: 20394608]

[110] Byrne PG, Simmons LW, Roberts JD. Sperm competition and the evolution of gamete morphology in frogs. Proc Biol Sci 2003; 270(1528): 2079-86. [http://dx.doi.org/10.1098/rspb.2003.2433] [PMID: 14561298]

[111] Kleven O, Laskemoen T, Fossøy F, Robertson RJ, Lifjeld JT. Intraspecific variation in sperm length is negatively related to sperm competition in passerine birds. Evolution 2008; 62(2): 494-9. [http://dx.doi.org/10.1111/j.1558-5646.2007.00287.x] [PMID: 18070085]

[112] Lifjeld JT, Laskemoen T, Kleven O, Albrecht T, Robertson RJ. Sperm length variation as a predictor of extrapair paternity in passerine birds. PLoS One 2010; 5(10): e13456. [http://dx.doi.org/10.1371/journal.pone.0013456] [PMID: 20976147]

[113] Lüpold S, Linz GM, Rivers JW, Westneat DF, Birkhead TR. Sperm competition selects beyond relative testes size in birds. Evolution 2009; 63(2): 391-402. [http://dx.doi.org/10.1111/j.1558-5646.2008.00571.x] [PMID: 19215291]

[114] Immler S, Birkhead TR. Sperm competition and sperm midpiece size: no consistent pattern in passerine birds. Proc Biol Sci 2007; 274(1609): 561-8. [http://dx.doi.org/10.1098/rspb.2006.3752] [PMID: 17476777]

[115] Balshine S, Leach BJ, Neat F, Werner NY, Montgomerie R. Sperm size of African cichlids in relation to sperm competition. Behav Ecol 2001; 12(6): 726-31. [http://dx.doi.org/10.1093/beheco/12.6.726]

[116] Gay L, Hosken DJ, Vasudev R, Tregenza T, Eady PE. Sperm competition and maternal effects differentially influence testis and sperm size in *Callosobruchus maculatus*. J Evol Biol 2009; 22(5): 1143-50. [http://dx.doi.org/10.1111/j.1420-9101.2009.01724.x] [PMID: 19309491]

[117] Morrow EH, Gage MJG. The evolution of sperm length in moths. Proc Biol Sci 2000; 267(1440): 307-13. [http://dx.doi.org/10.1098/rspb.2000.1001] [PMID: 10714886]

[118] Fitzpatrick JL, Baer B. Polyandry reduces sperm length variation in social insects. Evolution 2011; 65(10): 3006-12. [http://dx.doi.org/10.1111/j.1558-5646.2011.01343.x] [PMID: 21967440]

[119] Tourmente M, Gomendio M, Roldan ER. Sperm competition and the evolution of sperm design in mammals. BMC Evol Biol 2011; 11(1): 12. [http://dx.doi.org/10.1186/1471-2148-11-12] [PMID: 21232104]

[120] Tourmente M, Gomendio M, Roldan ER. Mass-specific metabolic rate and sperm competition determine sperm size in marsupial mammals. PLoS One 2011; 6(6): e21244. [http://dx.doi.org/10.1371/journal.pone.0021244] [PMID: 21731682]

[121] Anderson MJ, Nyholt J, Dixson AF. Sperm competition and the evolution of sperm midpiece volume in mammals. J Zool 2005; 267: 135-42. [http://dx.doi.org/10.1017/S0952836905007284]

[122] Tourmente M, Gomendio M, Roldan ER, Giojalas LC, Chiaraviglio M. Sperm competition and reproductive mode influence sperm dimensions and structure among snakes. Evolution 2009; 63(10): 2513-24. [http://dx.doi.org/10.1111/j.1558-5646.2009.00739.x] [PMID: 19490075]

[123] Bernard RTF. Reproduction of *Rhinolophus clivosus* (Microchiroptera) in Natal, South Africa. Z Saugetierkd 1982; 48(3): 321-9.

[124] Bernard RTF. Reproduction in the Cape horseshoe bat (*Rhinolophus capensis*) from South Africa. S Afr J Zool 1985; 20: 129-35. [http://dx.doi.org/10.1080/02541858.1985.11447925]

[125] Bernard RTF. The adaptive significance of reproductive delay phenomena in some South African Microchiroptera. Mammal Rev 1989; 19: 27-34. [http://dx.doi.org/10.1111/j.1365-2907.1989.tb00399.x]

[126] Rasweiler JJ IV. Prolonged receptivity to the male and the fate of spermatozoa in the female black mastiff bat, *Molossus ater*. J Reprod Fertil 1987; 79(2): 643-54. [http://dx.doi.org/10.1530/jrf.0.0790643] [PMID: 3646338]

[127] Crichton EG, Krutzsch PH. Reproductive biology of the female little mastiff bat, *Mormopterus planiceps* (Chiroptera: Molossidae) in southeast Australia. Am J Anat 1987; 178(4): 369-86. [http://dx.doi.org/10.1002/aja.1001780408] [PMID: 3604956]

[128] Lloyd BD. Advances in New Zealand mammalogy 1990–2000: short-tailed bats. J R Soc N Z 2001; 31(1): 59-81. [http://dx.doi.org/10.1080/03014223.2001.9517639]

[129] De Bonilla H, Rasweiler JJ IV. Breeding activity, preimplantation development, and oviduct histology of the short-tailed fruit bat, *Carollia*, in captivity. Anat Rec 1974; 179(3): 385-403. [http://dx.doi.org/10.1002/ar.1091790309] [PMID: 4837767]

[130] Rasweiler JJ IV, Badwaik NK. Delayed development in the short-tailed fruit bat, *Carollia*

perspicillata. J Reprod Fertil 1997; 109(1): 7-20.
[http://dx.doi.org/10.1530/jrf.0.1090007] [PMID: 9068408]

[131] Hernández GR, Herrera LG. Allocation of endogenous nutrients for reproduction in the lesser long-nosed bat (*Leptonycteris yerbabuenae*) in Central Mexico. J Mammal 2016; 97(1): 23-31.
[http://dx.doi.org/10.1093/jmammal/gyv149]

[132] Martin L, Kennedy JH, Little L, Luckhoff HC, O'Brien GM, Pow CST, *et al.* The reproductive biology of Australian flying-foxes (genus Pteropus). Symposia of the Zoological Society of London. 1995; pp. 167-84.

[133] Jones FW. The genitalia of the Cheiroptera. J Anat 1916; 51(1): 36-60.
[PMID: 17103802]

[134] Racey PA. The prolonged survival of spermatozoa in bats. The Biology of the Male Gamete 1975; 7(1): 385-416.

[135] Oh YK, Mōri T, Uchida TA. Studies on the vaginal plug of the Japanese greater horseshoe bat, *Rhinolophus ferrumequinum nippon*. J Reprod Fertil 1983; 68(2): 365-9.
[http://dx.doi.org/10.1530/jrf.0.0680365] [PMID: 6683318]

[136] Uchida TA, Mori T. Prolonged storage of spermatozoa in hibernating bats. In: Fenton MB, Racey P, Rayner JMV, Eds. Recent Advances in the Study of Bats. Cambridge, UK: Cambridge University Press 1987; pp. 351-65.

[137] Krutzsch PH. Anatomy, physiology and cyclicity of the male reproductive tract. Reproductive biology of bats. London, UK: Academic Press 2000; pp. 91-155.
[http://dx.doi.org/10.1016/B978-012195670-7/50005-9]

[138] Matthews LH. The form of the penis in the British rhinolophid bats, compared with that in some of the vespertilionid bats; and (2) The female sexual cycle in the British horse-shoe bats, *Rhinolophus ferrum-equinum insulanus* barrett-hamilton and *R. hipposideros minutus* montagu. Trans Zool Soc Lond 1937; 23(4): 213-23.
[http://dx.doi.org/10.1111/j.1096-3642.1937.tb00338.x]

[139] Gaisler J. Reproduction in the lesser horseshoe bat (*Rhinolophus hipposideros hipposideros* Bechstein, 1800). Bijdr Dierkd 1966; 36: 45-62.

[140] Krutzsch PH, Young RA, Crichton EG. Observations on the reproductive-biology and anatomy of rhinolophus-megaphyllus (chiroptera, rhinolophidae) in eastern Australia. Aust J Zool 1992; 40(5): 533-49.
[http://dx.doi.org/10.1071/ZO9920533]

[141] Kitchener DJ. Reproduction in female Gould's wattled bat, *Chalinolobus gouldii* (Gray) (Vespertilionidae), in Western Australia. Aust J Zool 1975; 23(1): 29-42.
[http://dx.doi.org/10.1071/ZO9750029] [PMID: 1140499]

[142] Kitchener DJ, Coster P. Reproduction in female *Chalinolobus morio* (Gray)(Vespertilionidae) in south-western Australia. Aust J Zool 1981; 29(2): 305-20.
[http://dx.doi.org/10.1071/ZO9810305]

[143] Pearson OP, Koford MR, Pearson AK. Reproduction of the lump-nosed bat (*Corynorhinus rafinesquei*) in California. J Mammal 1952; 33(3): 273-320.
[http://dx.doi.org/10.2307/1375769]

[144] Gates WH. Keeping bats in captivity. J Mammal 1936; 17(3): 268-73.
[http://dx.doi.org/10.2307/1374428]

[145] Wimsatt WA. An analysis of implantation in the bat, *Myotis lucifugus lucifugus*. Am J Anat 1944; 74(3): 355-411.
[http://dx.doi.org/10.1002/aja.1000740304]

[146] Kitchener DJ, Halse SA. Reproduction in female *Eptesicus regulus* (Thomas)(Vespertilionidae), in

south-western Australia. Aust J Zool 1978; 26(2): 257-67.
[http://dx.doi.org/10.1071/ZO9780257]

[147] Tidemann CR. Sex differences in seasonal changes of brown adipose tissue and activity of the Australian vespertilionid bat *Eptesicus vulturnus*. Aust J Zool 1982; 30(1): 15-22.
[http://dx.doi.org/10.1071/ZO9820015]

[148] Tidemann CR. Reproduction in the *Bats Vespadelus-Vulturnus*, *V-Regulus* and *V-Darlingtoni* (Microchiroptera, Vespertilionidae) in Coastal South-Eastern Australia. Aust J Zool 1993; 41: 21-35.
[http://dx.doi.org/10.1071/ZO9930021]

[149] Krutzsch PH, Crichton EG. Reproductive biology of the male bent-winged bat, *Miniopterus schreibersii* (Vespertilionidae) in southeast South Australia. Acta Anat 1990; 139(2): 109-25.
[http://dx.doi.org/10.1159/000146987] [PMID: 2288198]

[150] Thomas DW, Brock Fenton M, Barclay RM. Social behavior of the little brown bat, *Myotis lucifugus*. Behav Ecol Sociobiol 1979; 6(2): 129-36.
[http://dx.doi.org/10.1007/BF00292559]

[151] Wang Z, Liang B, Racey PA, Wang Y-L, Zhang S-Y. Sperm storage, delayed ovulation, and menstruation of the female Rickett's big-footed bat (*Myotis ricketti*). Zool Stud - Taipei 2008; 47(2): 215-21.

[152] Racey PA. The reproductive cycle in male noctule bats, *Nyctalus noctula*. J Reprod Fertil 1974; 41(1): 169-82.
[http://dx.doi.org/10.1530/jrf.0.0410169] [PMID: 4431008]

[153] Hosken DJ. Sperm fertility and skewed paternity during sperm competition in the Australian long-eared bat *Nyctophilus geoffroyi* (Chiroptera: Vespertilionidae). J Zool 1998; 245(1): 93-100.
[http://dx.doi.org/10.1111/j.1469-7998.1998.tb00076.x]

[154] Hiraiwa YK, Uchida TA. Fertilization in the bat, Pipistrellus abramus abramus (Temminck) : III Fertilization capacity of spermatozoa stored in the uterus after copulation in the fall. Sci B Fac Agr Kyushu 1956; 15(4): 565-74.

[155] Bernard RTF, Happold DCD, Happold M. Sperm storage in a seasonally reproducing African vespertilionid, the banana bat (*Pipistrellus nanus*) from Malawi. J Zool 1997; 241(1): 161-74.
[http://dx.doi.org/10.1111/j.1469-7998.1997.tb05506.x]

[156] Courrier R. Le cycle sexuel chez la femelle des mammifères: étude de la phase folliculaire. Imprimerie H. Vaillant-Carmanne 1924.

[157] Gerell R, Lundberg K. Social organization in the bat *Pipistrellus pipistrellus*. Behav Ecol Sociobiol 1985; 16(2): 177-84.
[http://dx.doi.org/10.1007/BF00295153]

[158] Strelkov PP. The peculiarities of reproduction in bats (Vespertilionidae) near the northern border of their distribution. Proceedings of the International Symposium on Methods in Mammalogical Investigation. Brno, Prague. 1962; pp. 306-11.

[159] Entwistle AC, Racey PA, Speakman JR. Social and population structure of a gleaning bat, *Plecotus auritus*. J Zool 2000; 252(1): 11-7.
[http://dx.doi.org/10.1111/j.1469-7998.2000.tb00815.x]

[160] Krishna A, Dominic CJ. Storage of spermatozoa in the female genital tract of the vespertilionid bat, *Scotophilus heathi*. J Reprod Fertil 1978; 54(2): 319-21.
[http://dx.doi.org/10.1530/jrf.0.0540319] [PMID: 569203]

[161] Orr T,J, Brennan PLR. All features great and small-the potential roles of the baculum and penile spines in mammals. Integrative and Comparative Biol 2016; 56(4): 635-43.

GLOSSARY

Acrosomal reaction Fusion between the sperm plasma membrane and the outer membrane of the acrosome releasing the contents of it to the external environment and leaving the sperm acrosome protected by the inner membrane, which help the sperm to move through the zona pellucida, to fertilize the egg.

Acrosome Small tank located at the apical end of the sperm head and containing hydrolytic enzymes.

Androgens Steroid hormones (derived from cyclopentanoperhydrophenanthrene) whose main function is to stimulate the development of male sexual characteristics.

Anisogamia Form of sexual reproduction in which the bind male and female gametes are morphologically different.

Annulus A structure located at the distal end of the midpiece of the sperm flagellum; the annulus is thought to form a diffusion barrier between the midpiece and the principal piece and serve as a stabilizing structure for tail rigidity.

Apocrine Ejecting gland secretory product released together with a part of the cell.

Arousal Physiological and psychological state of being awake.

Autocrine Cell signaling in which a cell secretes a hormone or chemical messenger (called the autocrine agent) that binds to autocrine receptors on that same cell leading to changes in the cell.

Axoneme An axial inner structure of cilia and flagella of eukaryotic microtubular basically which is the essential element for mobility.

Baculum A bone found in the penis of most mammals more precisely in insectivores, rodents, carnivores and primates. Among domestic species, it is found in the cat and dog. This bone allows these animals penetration in the absence of erection.

Breeding The reproduction that is producing of offspring usually animals or plants.

Canicula Season of the year in which heat is stronger both in the South and the North.

Chromatin The way that presents the DNA in the cell nucleus. It is the substance of eukaryotic chromosomes base corresponding to the association of DNA, RNA and proteins found in the interphase nucleus of eukaryotic cells and constitutes the genome.

Cumulus cells Cells surrounding the both oocyte and the ovarian follicle after ovulation. It´s functions are oocyte protection, coordination of follicular development and oocyte maturation.

Cunnilingus It is the practice of oral sex in the female genitals.

Cytokinesis The process during cell division in which the cytoplasm of a single eukaryotic cell is divided to form two daughter cells.

Diverticulae From the medical or biological term for an outpouching of a hollow (or a fluid-filled) structure in the body.

Ensiform Shaped sword structure.

Estromal cell Connective tissue cells of any organ.

Euchromatin Lightly packed form of chromatin (heterochromatin unless that is often not active almost never) with a high concentration of genes, and often (not always) is in active transcription.

Edith Arenas-Ríos (Ed.)

Excoriation Skin irritation that occurs where skin rubs against itself clothes or other material.

Falciform Sickle form structure.

Fellatio It is a form of oral sex is sucking or licking the penis or scrotum with the mouth lips, or tongue.

Glans The sensitive bulbous structure at the distal end of the penis.

Harem Of Arab means harem which forbidden, this term is used to mean women sanctuary or more women.

Heterothermic It applies to animals that can not regulate its temperature and therefore have the same of the living environment.

Hypertrophy Growth or increase in size of muscle fibers.

Horn Point where the uterus and the fallopian tube meet.

Inguinal The inguinal canal in adults is the oblique passage of about 4 cm long, inferomedially directed through the lower portion of the anterolateral abdominal wall.

Intrasexual Between individuals of the same sex.

Lectins proteins that bind to sugars with high specificity for each different type. Its main role is in recognition phenomena both molecular and cellular levels.

Manchette A transient skirt-like structure surrounding the elongating spermatid head and is only present during spermatid elongation.

Mating It is the set of all courtship and breeding behaviors that perform two individuals of different sex to procreate and culminates in copulation.

Merocrine Mode widespread secretion from exocrine cells (glands that secrete substances ejected through an excretory channel). In the case of a secretory merocrine cell does not deteriorate because of the synthesized substance release is by exocytosis.

Metabolites Intermediates and products of metabolism. They have various functions including fuel, structure, signaling, stimulatory and inhibitory effects on enzymes, catalytic activity of their own (usually as a cofactor to an enzyme), defense, and interactions with other organisms.

Paracrine Form of cell-cell communication which produces a signal to induce changes in nearby cells altering the behavior or differentiation of those cells.

Pearson correlation index Measure of the linear relationship between two random variables quantitative.

Pellucid zone Outer layer surrounding the mammals oocyte, separating the perivitelline space. It is composed of several glycoproteins, grouped into three families ZP1, ZP3 and ZP2. The glycoproteins have receptors that are activators for sperm, making egg fertilization possible.

Perforatorium Equivalent to the subacrosomal space and contents described by others in a variety of mammalian sperm.

Phenology Science that studies the relationship between climatic factors and cycles of living things.

Phylostomidae The more varied and diverse family from the order Chiroptera very diverse in their morphology and behavior. Most are insectivores, but there are many species nectarivorous and frugivorous.

Phytophagous An herbivore is an animal that feeds mainly on plants.

Piscivorous A carnivorous animal that feeds mainly on fish.

Poikilothermic Animal whose body temperature varies with the environment because it lacks regulatory mechanisms of the same all animals except birds and mammals.

Quiescence Cell cycle out and kept in phase G. This cell can return to mitotic process if need tissue regeneration.

Spermathecae An organ of the reproductive tract of the females of certain insects mollusks and certain other invertebrates and vertebrados.1 Its purpose is to receive and store sperm received from the male. Generally that is where fertilization takes place, ie the union of the male female gamete when the oocyte is mature

Synaptonemal complexes Proteins structures that forms between homologous chromosomes (two pairs of sister chromatids) during meiosis and is thought to mediate chromosome pairing synapsis, and recombination.

Taxa A group of related organisms which in a given classification have been grouped and assigned a Latin name, according to their description, species, and a type.

Tight bonds Cytological structures present in the epithelium and endothelium cells that create a barrier preventing the free flow impermeability of substances between cells.

Torpid Who or what experience lethargy. Rest lack of activity, numbness, drowsiness or sleepiness.

Torpor Characterized state by a large decrease in the physiological activity in an animal usually evidenced by a reduced of body temperature and metabolic rate. A time of torpor may refer to the period of time a hibernator goes low body temperature, which can last days or weeks, or it can refer to a period of low body temperature and metabolism that lasts less than 24 hours, as in the “daily torpor”.

SUBJECT INDEX

A

Accessory 20, 24, 27, 51, 56, 76, 91
  glands, reproductive 20, 24, 27
  sex glands 51, 56, 76, 91
Acrosomal 9, 10, 54, 81, 87, 91
  cap 9, 10, 54
  reaction 81, 87, 91
Acrosome 10, 74, 109
  phase 10
  reaction 74, 109
Adenohypophysis 87
Adluminal compartment 3, 4, 5, 65
Amounts of ROS 90, 91
Androgen-binding protein (ABP) 87, 88, 107, 111
Androgen receptor (AR) 20, 22, 23, 24, 26, 30, 31, 61, 89, 107, 110, 111, 112
Angiogenesis 58, 60, 63, 64, 65
Annual reproductive cycle 29, 30, 51, 74, 89
Apical 80, 82
  cell 80
  section 82
Apocrine secretion 82
Apoptotic cells 65, 66
Aposomes 83
AR 22, 23, 25, 26, 30
  expression 22, 25, 26
  in Sertoli cells 23, 30
  of epididymal cells 24, 25
Aromatase 23, 24, 25, 26
  enzyme 24, 25
  expression 23, 25, 26
Axial filament 11, 12, 13
Axoneme 9, 10, 11, 82

B

Baculum 134, 135
  length 134, 135
Bare dorsal patch (BDP) 40, 41, 42, 43, 46
Barrier 54, 55, 80
  hemato-epididymal 80
  hematotesticular 54, 55
Basal lamina 3, 4, 5, 14, 15
Bats 12, 20, 21, 23, 26, 28, 31, 37, 49, 50, 51, 55, 59, 63, 65, 74, 81, 88, 104, 106, 112, 122, 126, 129, 132, 133, 134, 135, 136, 137
  cold zones 55
  frugivorous 50
  immature 59
  insectivorous 55
  lesser long-nosed 132
  male 20, 21, 28, 50, 51, 74, 129, 134
  mexicanus 49, 63, 65
  myotis nigricans 23, 26, 28
  nocturnal nonpteropodid 136
  non-seasonal polyestrous 88
  penises 135
  phyllostomid 37
  pipistrellus 106
  reproduction 31, 122, 137
  rhinolophid 104, 106
  seasonal Artibeus lituratus 81
  short-tailed 132
  short-tailed fruit 129, 132
  specialized hematophagous 50
  spermatozoon of 12
  sperm-storing 112
  temperate 28, 126
  temperate zone molossid 104
  tropical 104
  vespertilionid 59, 104
  wattled 133
Blastocysts 124
Blocks sperm transport 131
Blood vessels 49, 53, 56, 60, 61, 62, 64, 65, 66
  extratesticular 65, 66
  intratesticular 65, 66
  proportion of 60
Breeding 27, 29, 30, 36, 38, 43, 45, 46
  colony 36, 38, 43, 45, 46
  periods 27, 29, 30

C

Caput epididymis 24, 25
Carbohydrates 85, 86, 89, 92, 108

Edith Arenas-Ríos (Ed.)

Carnitine transporter 110, 111
Cauda epididymis 25, 26
Caudal region 74, 75, 78, 81, 84, 91, 92
Cauda regions 88, 89
Cell death 24, 25, 56, 57, 59, 64, 87
Cell morphology, epithelial 24, 25
Cells 1, 2, 4, 21, 24, 25, 26, 53, 56, 57, 59, 62, 63, 64, 65, 74, 77, 80, 105
  interstitial epididymal 24, 25
  mesodermal 62, 63
  stromal 25, 26
  testicular 21, 24
Cells release 3, 4
Cell types 3, 4, 5, 14, 21, 28, 56, 80, 85, 86, 87, 90
  diverse 80, 85
Cellular 1, 2, 63
  division 1, 2
  specificity 63
Centriolar adjunct 9, 11
Cervical ganglion, upper 52
Chiroptera 1, 16, 20, 23, 32, 50, 74, 119
Chromatin 3, 4, 6, 52, 62
  condensation 3, 6
  decondensed 4
  nuclear 54, 62
Chromosomes 6
Coelomic domain 63
Column, longitudinal 12, 13
Comparative analyses 65, 129, 134, 135
Copulatory plugs 130, 131, 132, 137
Corpus epididymis 25
Corynorhinus mexicanus bat 59, 78, 79, 83, 86, 87, 90, 91, 92
Cryptic choice 120, 121, 122, 136, 137
Cunnilingus 136
Cytoplasm 4, 5, 14, 15, 25, 54, 84
Cytoplasmic droplet 81, 91

D

Decreased sperm motility 112
Delayed 122, 123, 124, 126, 129, 132, 135
  development 123, 124, 132
  fertilization 122, 123, 124, 126, 129, 132, 135
  implantation 123, 124, 132
  ovulation re-sperm storage 135
Diakinesis 6
Digital Vernier 38
Dihydrotestosterone 22, 25, 26, 55, 88
Diplotene 6
Dorsal 26, 36, 37, 38, 40, 41, 42, 43, 45, 46
  patch (DP) 36, 37, 38, 40, 41, 42, 43, 45, 46
  prostate 26
Dry seasons 38, 44, 45

E

Electron-dense granules 8, 9
Electron-lucid 8, 9
  matrix 8, 9
Electron micrographs 6, 7, 8, 9, 10, 11, 15, 22
Embryonic development 62, 63
Endoplasmic reticulum 14, 15
Endothelial 63, 64
  cells 63, 64
  mesonephric cells 63
Endothelium 65
Energy 36, 37, 44, 45, 46
  expenditure 45
  reserve 36, 37, 44, 46
Epididimosomas 82
Epididymal cauda 79
Epididymal cells 24, 25
Epididymal duct 74, 85
Epididymal epithelium 31, 80
Epididymal functionality 93
Epididymal functions 88, 93
Epididymal maturation 74
Epididymal morphophysiology 30
Epididymal Regulation 24
Epididymal Sperm Maturation (ESM) 74, 75, 78, 83, 85, 86, 87, 90, 91, 92
Epididymal sperm storage 51, 77
Epididymal sperm storage capacities 44
Epididymal tubule 78, 79, 80, 82, 83, 87, 89
Epididymal tubule form 80
Epididymis experiences 30
Epididymisomes 81, 83, 93
Epithelial cells 24, 25, 26, 75, 82, 85, 87, 104, 105, 106, 107, 110
Epithelial cells cytoplasm 25, 106
Epithelial lining 104, 105, 107

ERα expression 23, 26
ERα expression in corpus epididymis 25
Evolutionary consequences of sperm competition 125

F

Fat Index (FI) 38, 40, 41
Fat pattern 36
Female genitals 135
Female genital tract 74, 103, 104, 106, 107
Female immune attack 131
Female mate-choice 120
Female remating 131
Female reproductive tract 75, 103, 105, 109, 110, 111, 112, 126, 128, 129, 131
Females sperm storage 137
Females storing sperm 107
Female tract epithelium 109
Final sperm production 21
Fluid 62, 79, 85, 86, 107, 108, 125, 130, 131, 134
  epididymal 85, 86
  seminal 107, 108, 130, 131
Fluid reabsorption 24, 25
Flutamide 107
Follicle-stimulating hormone (FSH) 20, 21, 22, 24, 27, 31, 54, 57, 58, 87
Forearm length 38
Form spermatozoids 54

G

Gene expression 75, 80, 87, 88
Genital crests 62, 63
Genital morphology 125, 134
Germ cells 20, 21, 23, 30, 59, 62, 63
  primordial 62, 63
Germinal cells 54, 57, 59, 61
Glycoconjugates 85
Glycolysis 107, 108, 126
Glycoproteins 84, 85, 87, 90
Glycosylation 84, 85, 91
Golgi complex 1, 7, 8, 9, 14
Granular euchromatin 15

H

Haploid cells, specialized 1, 2
Heat shock protein (HSPs) 109, 112
Helping sperm surve 131
Hemangioblasts 63
Hematophagous species desmodus rotundus 50
Heterosexual intercourses 43
Hibernating animals 28, 29
Hibernating bats 28, 29, 31, 78, 106, 109, 122
Hibernation periods 20, 27, 29, 106
Hormonal control of reproduction 27, 28
Hormonal regulation 20, 22, 26, 28, 32, 89
  pattern of 20, 28
Hormonal regulation of bats 20, 26, 27
Hormonal regulation of sperm-storage 107
Hormonal regulation of testicular regression in hibernating bats 28
Hormonal regulation of testicular regression in nonhibernating bats 29
Hormonal regulation patterns 29, 89
Hormones 20, 21, 22, 23, 27, 28, 29, 31, 52, 56, 57, 60, 87, 108, 110
  follicle-stimulating 20, 21, 22, 57, 87
  gonadotropin-releasing 20, 21, 52
  luteinizing 20, 21, 22, 52, 57, 87
Hormones act 20, 21
Hormone sensitive lipase (HSL) 108, 110, 111, 112
Horseshoe bats 133
Hypothalamic-pituitary-testicular axis 20, 21, 22

I

Increased androgen production 110
Initiating sperm motility 90
Inner acrosomal membrane 12, 13

L

Large cytoplasm 4
Leptonycteris Yerbabuenae 36, 37, 39, 132
Leptotene 6

Lesser horseshoe bats 133
Leydig cell cytoplasm 15
Leydig cells 4, 14, 15, 21, 22, 23, 29, 53, 54, 56, 57, 61, 77, 87, 88
LH levels 28, 29
Likelihood females favour 136
Lipid droplets 15, 16
Lipid peroxidation 53, 89
Lipids 84, 85, 89, 105, 108, 112
Long-eared bat species 51
Longitudinal microtubule manchette 10, 11
Long-term female sperm storage 131, 132
Long-term sperm storage 131, 132
Long-term sperm survival 131
Low body temperature facilitate sperm-storage 106
Lumen 14, 31, 53, 54, 61, 64, 65, 66, 81, 83, 87, 88, 89
  epididymal 85
Luminal microenvironments 75, 79, 80
Luteinizing hormone (LH) 20, 21, 22, 24, 27, 28, 29, 31, 52, 57, 58, 87

M

Male leptonycteris yerbabuenae 42, 43
Male reproductive functions and prolonged sperm storage 77
Male reproductive patterns 76
Male reproductive periodicity in mammals 75
Males 40, 41, 45, 122, 127, 134
  rival 127, 134
  temperate species 122
  yerbabuenae 40, 41, 45
Mammalian hormonal regulation 21
Mammals 1, 2, 13, 14, 49, 50, 53, 58, 74, 75, 76, 78, 85, 92, 103, 119, 134
  hibernating 76
  non-chiroptera 76, 92
Mammal species, non-seasonal 79
Mammals sperm 128
Mammals sperm shape 128
Mammal testicles 55
Mating system dependant 123, 124, 125
Meiotic activity 61, 62
Meiotic divisions 1, 2, 3, 4, 7, 53
Melatonin 31, 52
Mexican bat corynorhinus 79, 80
Mexicanus 51, 59, 60, 65
Migration and reproduction 36
Migration of endothelial cells 63
Mitotic divisions 3, 4
Modification of sperm proteins during maturation 84
Modulating sperm motility 83
Molossidae 1, 14, 55, 132
Molossus molossus 5, 6, 7, 8, 9, 11, 60
Monoestrous 28, 29, 91
Morphology, spermatozoon head 13
Multimale-multifemale mating system 134
Myotis lucifugus 81, 104, 106, 107, 122

N

N-acetyl glucosamine 86, 92
Neotropical species molossus 14
Neotropical species of bats 7
New sperm production 51
Nonhibernating bats 29
  subtropical 29
Non-hormonal factors 87, 88
Nonpalpable testes 38, 39
Non-spermatogenic state 130
Northwest 30
Nuclear elongation 10, 11
Nucleus 5, 6, 7, 8, 9, 10, 11, 12, 13, 14, 15, 54, 59, 75
Nutrients 54, 60, 62, 105, 107, 112
Nutritional support 107, 108

O

Outer acrosomal membrane 12, 13
Outer dense fibers 1, 12, 13, 14
  larger 12
Oviduct length 135
Ovulation, induced 123, 124

P

Pachytene 6, 59
Paracrine communication 20, 21
Paraguay 30
Participation of apoptosis 49, 56, 57, 65

Participation of epididymisomes 81
Paternal genes 124
Penis morphology 134
Perforatorium 1, 9, 11, 12, 13
Pericytes 64
Phosphatases 86, 90
Phosphorylation 86, 91
Photoperiod 52, 53, 58
Photosensitive species– 58
Phyllostomidae 1, 14, 37, 132
Physiological processes 89, 90
Physiology, reproductive 52, 76
Pineal gland 52
Pipistrellus 77, 107, 134
Pituitary gland 20, 21
Plasma membrane 9, 12, 13, 109
Plasmatic membrane 81, 83, 84, 85
Platyrrhinus lineatus 5, 6, 8, 9, 10, 11, 14
Plug Origin 132, 133
Polyandry 120, 122
Post-copulatory 122, 123, 125, 131, 135, 136, 137
Pre-copulatory cunnilingus 136, 137
Proacrosomal vacuole 7, 8, 9, 10
Proacrosomal vesicles 8, 9
Proacrosomal vesicles and vacuole 8, 9
Process, spermatogonial differentiation 3, 4
Production, female gamete 109
Prolonged sperm storage 26, 77, 78, 81, 83, 103, 107
Prolonged sperm storage process 78, 105
Prolonged storage of viable spermatozoa 106
Prolonged storage sperm 74
Protein expression 24, 25
Protein glycosylation 84, 86
Protein phosphorylation 86, 90, 91, 92
Proteins 24, 25, 74, 75, 81, 82, 83, 84, 85, 86, 87, 88, 90, 91, 106, 107, 108, 110, 111, 131
  androgen binding 88, 107, 110, 111
  incorporation of 81
  kinase 86, 90
  transfer 83
Protein synthesis 75, 80, 93

R

Reactive oxygen species (ROS) 53, 74, 89, 90, 91, 92, 93
Region 11, 13, 14, 15, 38, 50, 75, 79, 80, 81, 83, 87, 93, 104, 127, 135
  apical 14, 15
  basal 14, 15
  bats storage 127
  epididymal 80
  polyestrous bats inhabit tropical 50
  rostral 11, 13
  specialized storage 104
Regression 30, 64, 129
  interspecific 129
Regulation 21, 22, 23, 25, 26, 58, 87, 88
  endocrine 25, 26
Regulation by reactive oxygen species 89
Relationship, intimate 103, 106
Release sperm 111
Reproductive adaptations 27, 32
Reproductive behaviors 44, 52
Reproductive functions, secondary 55, 91
Reproductive inactivity 58, 59, 65
Reproductive organs 20, 21, 31, 62
Reproductive patterns 49, 51, 55, 77
  asynchronous 49, 55
  asynchronous male 51, 77
Reproductive phenology 38, 39
Reproductive process 16, 20, 21, 28
Reproductive seasons 52, 80
Reproductive success 25, 32, 36, 46, 120, 131
Reproductive tracts 106, 120, 121, 122
ROS production 89, 90

S

Scotophilus heathi 55, 104, 105, 106, 107, 109
Scrotal especies 59
Scrotal testes 38, 39, 44
SC-sperm competition 123
Seasonal spermatogenesis 76
Seasonal variation in spermatogenesis 55
Sebaceous 40, 42, 43

Sebaceous dorsal patch (SDP) 40, 42, 43
Secondary reproductive organs 24, 27, 29
Secondary sex organ 76, 77
Secondary sex structure 20, 21
Secretor granules 83
Section thickness 62
Seminal receptacle 121
Seminiferous epithelium 3, 28, 54, 56, 57, 59, 61, 62
Seminiferous tubules 3, 4, 14, 15, 20, 21, 53, 54, 55, 56, 61, 62, 65, 66, 87
Sertoli cells 3, 5, 14, 15, 21, 22, 23, 24, 28, 29, 30, 54, 56, 57, 59, 61
  inactive 28
Sexual conflict 119
Sexual dimorphism 119
Sialic acid 86, 92
Significance of sperm storage 112
Single sperm shape 129
Site of sperm storage 104
Somatic cells, testicular 20, 21
Species 41, 45, 49, 50, 51, 52, 55, 56, 58, 60, 64, 77, 79, 88, 89, 91, 92, 104, 125, 126, 127, 129, 130, 131, 132, 134, 137
  endemic 51, 91
  hibernating 55
  insectivorous 51
  mammalian 45, 129, 131
  monoestrous 50, 79
  monogamous 126, 132, 134
  non-storing 137
  plecotine 60
  pollinivorous 50
  polyestrous 50
  reproducing bat 79
  reproductive bat 49, 64
  seasonal bat 56
  tropical bat 60
  vertebrate 104
  antrouzous pallidus 52
  antrozous pallidus 58
  eptesicus fuscus 77
  level 134
  mating 130, 131
  mating season 41
  myotis nigricans 89
  myotis ricketti 88
  -rich mammalian order 125
  sperms 92
  sperm size 127
Sperm 13, 44, 56, 75, 79, 83, 86, 87, 90, 92, 107, 108, 109, 111, 112, 120, 121, 123, 126, 128, 129, 130, 131, 135, 137
  buffer 131
  capacitated 109
  co-culture 112
  competing 120, 121
  defective 83
  dump 131
  epididymal 86, 90
  fertile 83
  green 92
  healthier 108
  mammalian 126
  mature 44, 56, 75, 79, 86, 87, 92
  non-preferred 121
  particular 137
  slower 129
  stored 83, 107, 108, 111, 123, 135
  storing 108
  streamlined 128
  transfer 130
  trapped 135
  typical bat 13
  viable 137
Sperm approach maturity 86
Spermathecae 104, 121
Spermatic cord 62
Spermatic production 29, 30
  large 29
Spermatid differentiation 5, 7
Spermatids 1, 2, 4, 5, 7, 8, 9, 10, 11, 14, 15, 22, 23, 24, 30, 49, 53, 54, 57, 59
  elongated 24, 30
  haploid 1, 2, 4
  round 1, 2, 5, 7, 54, 59
Spermatids of Molossus molossus 7, 11
Spermatids of Platyrrhinus lineatus 9, 10
Spermatid types 62
Spermatocytes 1, 2, 3, 4, 14, 15, 49, 53, 57, 59, 61
  primary 53, 59, 61
  producing 2
  secondary 53, 57
  spontaneous degeneration of 49, 57

Spermatogenesis 2, 5, 21, 55, 56, 61, 62, 65, 77, 92, 104, 127, 137
  cessation of 104, 137
  complete 65
  continuous 127
  early stage of 61, 62
  longest phase of 2, 5
  normal 21
  period of 55, 56, 77, 92
Spermatogenesis process 2, 14, 16, 20, 21, 22, 29, 56
Spermatogenesis-regulating factors 55
Spermatogenic activity 30, 50, 52, 65, 66, 76
  peaks of 30
Spermatogenic production 27, 29
  intense 29
Spermatogonia 1, 2, 3, 4, 5, 24, 28, 30, 49, 53, 54, 57, 58, 59, 61, 63
  intermediate 57
  main types of 1, 4
  proliferation of 49, 57, 58, 59
  type Ap 3, 4, 5
Spermatogonial differentiation 1, 21
Spermatogonial stem cell 3, 4
Spermatogonia of platyrrhinus lineatus 5
Spermatogonia proliferate 1
Spermatogonia recruitment, normal 54
Spermatozoa 1, 2, 5, 11, 12, 13, 24, 25, 27, 30, 31, 92, 103, 104, 105, 106, 107, 108, 109, 112
  basic bat 12
  boar 109
  immature 92
  intact 104
  luminal 24, 25
  mature 1, 2, 5, 92
  production of 27
  stored 107, 108, 109, 112
  storing 107
Spermatozoa and storage organ 106
Spermatozoa being 31
Spermatozoid 54
Spermatozoid production 54
Spermatozoon head 12, 14
Spermatozoon surface 11
Sperm attributes 137
Sperm being 78
Sperm binding 109
Sperm biologists 129
Sperm biology 137
Sperm cells 62, 63, 74, 91, 92, 112
  turquoise 92
Sperm characteristics 127
Sperm competition 119, 120, 121, 122, 125, 126, 127, 128, 129, 131, 132, 134, 135, 136, 137
  disentangle 136
  intense 126
  outcome of 121, 131
Sperm competition and sperm numbers 125
Sperm competition raffle 121
Sperm competition risk 125, 126, 127, 129, 131, 134, 135
  increased 126
  indicators of 126
Sperm competition success 121
Sperm competition theory 125, 126, 130
Sperm DNA-repair mechanisms 112
Sperm dumping 124, 132
Sperm function 131, 137
Sperm head 5, 106, 109, 128, 129
  decondensing bat 109
Spermiation 12, 54
  process 12
Sperm integrity 107
Spermiogenesis 1, 2, 5, 7, 9, 10, 11, 22, 54
  process 5, 7, 9, 10, 11
Sperm leakage 131, 132
Sperm length 127, 128, 129
Sperm length variation in length 128
Sperm longevity 105, 109, 123
  prolonged 105, 109
Sperm loss, preventing 111
Sperm maintenance 80, 125
Sperm males transfer 121
Sperm maturation 44, 51, 74, 78, 79, 82, 85, 89, 90, 92, 93, 131
  final 78
  inducing 79
  process of epididymal 74, 92
  propitiates 89
Sperm membrane 83, 84, 92
Sperm membrane remodeling 24, 25
Sperm metabolism 85, 107, 108
  suppress 107, 108
Sperm metrics 129

Sperm morphology 125, 127, 129
Sperm morphometry 126, 129
Sperm motility 24, 25, 104, 106, 126
Sperm numbers 125, 126, 129, 136
Spermophilus citelus and mesocricetus auratus 76
Sperm performance 135
Sperm phenotype 124, 137
Sperm physiology 90
Sperm plasma membrane 90
Sperm production 27, 53, 76, 78, 125, 126, 137
Sperm proteins during maturation 84
Sperm's flagella 86
Sperm's nucleus 126
Sperm- storage 110
Sperm storage area 76
Sperm-storage crypts, developed 104
Sperm storage in female genital tract 104
Sperm storage mechanism 112
Sperm-storage organs 105, 121
Sperm-storage phenomenon 106
Sperm-storage process 108, 112
  natural 108
Sperm-storage site 104, 108, 112
Sperm storage sites 131
Sperm-storage tubules 104
Sperm-storing 123
Sperms traverse 79
Sperm survival 104, 109, 112
  permit 104
  prolonged 109, 112
Sperm surviving 112, 131
Sperm trait relationship 127, 128
Sperm traits 121, 129
Sperm transit 75
Sperm viability 78, 89, 109, 112
  increased 109
  maintaining 78
Spines, penile 135
Stages 65, 123, 124
  reproductive 123, 124
  spermatogenic 65
Stem cell pool 1, 2, 3, 4
Steroid hormones 20, 21, 57
Stimulate/regulate epididymal cells 31
Stimulate sperm dumping 135
Stimulation of apoptosis in germinal cells 57
Storage organs 105, 106, 107, 127
  female 105, 106
  female sperm 127
Storage site 104, 107, 111, 131
Storage sperm 74
Storage sperm period 92
Stores 120, 137
  female sperm 137
  large sperm 120
Subcellular structure 14, 15
Subcutaneous fat 36, 44
  males lack 44
Superfoetation 123, 124, 125
Supporting cells of bats 14
Suppressed sperm motility 108
Synaptonemal complexes 4, 6

T

Taphozous georgianus, emballonurid bats 27
Taxonomic groups 103, 105
Temperate 51, 55, 65, 122
Temperate regions 76, 106
Temperate seasonal species 60
Temperate species 122, 127
Temperate zone bats 30, 52, 56, 75, 104
  hibernating 104
Temporal asynchrony 60, 77, 78, 91
Testicles 49, 52, 53, 55, 56, 57, 58, 59, 62, 63, 65, 75, 78, 87, 88
Testicles of mammals 58
Testicular activity 27, 31, 60, 75, 76, 77, 88
Testicular arteries 60, 61, 63, 65, 66
Testicular cord 63
Testicular cycle 27, 49, 59, 61, 65
Testicular development 49, 57, 77
Testicular factors 30, 87, 88
Testicular functions 23, 24, 60
Testicular inactivity 59, 60
Testicular involution 58, 62, 75, 88
Testicular mass 36, 38, 39, 40, 41, 42, 43, 44, 45, 46, 56, 58
Testicular production 29, 30, 31
Testicular reactivation 29, 30
Testicular regression 29, 30, 31
  periods of 30
  process of 29, 31

Testicular steroids 20, 21
Testicular testosterone, action of 25, 26
Testicular vasculature 60
Testis mass 129, 135
Testis size 44, 125, 126, 129, 130
Testosterone 20, 21, 22, 25, 26, 54, 55, 57, 58, 87, 88, 89, 107, 108
Testosterone levels 55
Testosterone synthesis 87
Tight junctions 54
Tissues 15, 62, 77, 80, 81, 82, 83, 85, 108
  epididymal 82
  epithelial 80, 81, 83, 85
  interstitial 15, 62, 77
  sperm storing 108
Tools, species identification 134
Total flagellum length head length 128
Total sperm length 128
Tract morphology, female reproductive 128, 129
Tropical species 60, 77, 78, 122
  nonhibernating 78
Typical mammalian sperm model 13
Tyrosine phosphorylation 86, 87, 90, 91, 93
Tzinacanostoc populations males 45

U

Ultrastructure of bat spermatozoa 12
Uterine sperm motility 109
Utero-tubal junction (UTJ) 104, 105, 106, 107, 108, 109, 110, 112

V

Variation in sperm length 127
Variations, sperm size 129
Vascular elements 60, 61
Vascular plexus 63
Vasculogenesis 63
Veins 61, 62, 63, 64
  testicular 61
Velifer 77, 106, 107, 131
Vespertilionid, female 106
Vespertilionidae 1, 13, 14, 23, 55, 74, 76, 77, 133
Vessels, intra-testicular 62
Viable spermatozoa 106
Volume 38, 135
  sperm mid-piece 135
  testicular 38

W

Weaponry 119
Winter dormancy 31, 76, 106
Without dorsal patch (WDP) 40, 41, 42, 43

Y

Yerbabuenae 37, 38, 39, 42, 44, 45, 46

www.ingramcontent.com/pod-product-compliance
Lightning Source LLC
LaVergne TN
LVHW070952210926
848482LV00002B/192
* 9 7 8 1 6 8 1 0 8 5 5 5 5 *